Advances in Turfgrass Science

Management of Turfgrass Diseases

SECOND EDITION

by

J. M. Vargas, Jr.

CRC Press
Taylor & Francis Group
Boca Raton London New York

CRC Press is an imprint of the
Taylor & Francis Group, an **informa** business
A TAYLOR & FRANCIS BOOK

CRC Press
Taylor & Francis Group
6000 Broken Sound Parkway NW, Suite 300
Boca Raton, FL 33487-2742

First issued in paperback 2019

ISBN-13: 978-1-56670-046-7 (hbk)
ISBN-13: 978-0-367-40219-8 (pbk)
Library of Congress Card Number 93-28186

Library of Congress Cataloging-in-Publication Data

Vargas, J. M., Jr.
Management of turfgrass diseases / J.M. Vargas, Jr.–[2nd. ed.]
p. cm.
Includes bibliographical references (p.) and index.
ISBN 1-56670-046-9
1. Turfgrasses—Diseases. 2. Turf management. I. Title.
SB608.T87V37 1994
635.9'64293—dc20 93-28186
CIP

Visit the Taylor & Francis Web site at
http://www.taylorandfrancis.com

and the CRC Press Web site at
http://www.crcpress.com

To Rita and Joe

My Mother and Father

Thanks for everything

J.M. Vargas, Jr., has been a professor of Botany and Plant Pathology for the past 25 years at Michigan State University, where he has been involved in teaching, research, and extension. He received his BS degree from the University of Rhode Island in 1963, his MS from Oklahoma State University in 1965, and his PhD from the University of Minnesota in 1968. Dr. Vargas has published over 200 articles on turfgrass diseases and related subjects and has given over 1,000 presentations at conferences throughout the world, including Australia, New Zealand, South Africa, Japan, and Europe. His research accomplishments include:

- discovering the first bacterial disease in turf on Toronto creeping bentgrass after many scientists failed to discover the problem;
- reporting on the first resistance by a turfgrass pathogen to a systemic fungicide and later reporting the first resistance to the DMI fungicide by the dollar spot organism;
- identifying the true cause of the black layer as excess sulfur in the soil;
- demonstrating that annual bluegrass does not die from high temperature in cool season grass growing areas, but from two diseases, anthracnose and summer patch;
- developing the first mathematical prediction model for a turfgrass disease;
- developing a fungicide timing model for summer patch that is used throughout the world;
- helping to develop the first biological control for annual bluegrass.

Series Preface

Turfgrass science is comparatively young compared to most areas of applied plant science. The first revolutionary breakthrough was the invention of the reel mower in 1830. Then in 1906 the first book, entitled *Lawns and How to Make Them*, was published by author Leonard Barron. This was followed by a number of books published in the next six decades that were based on the art of turfgrass culture, developed by trial-and-error methods for the most part. A small pool of basic knowledge of turfgrass science was developed through research in the 1920s and then was greatly extended after 1950. The publication of *Turfgrass: Science and Culture* in 1973 documented a new era in the science of turfgrass culture. The generation of research information and technical developments in turfgrass science has continued to accelerate even more rapidly since then.

Now there is a need for yet more specific science-based books that encompass a practitioner orientation for the individual technical areas within the turfgrass science and culture fields. Accordingly, this ADVANCES IN TURFGRASS SCIENCE series has been developed as a coordinated, well-planned series of books encompassing the major areas of turfgrass science, culture, IPM, management, construction, and equipment.

This book, *Management of Turfgrass Diseases*, by Dr. Joseph M. Vargas, Jr., addresses a major aspect of turfgrass science. It offers a science-based technical presentation for both the student and the practitioner interested in the diseases of turfgrasses and their management. A key strength of the book is the author's integration of the individual turfgrass disease discussions with real-world cultural conditions.

—Dr. James B. Beard

Preface

This book is written primarily for undergraduate students majoring in turf and turfgrass managers; i.e., golf course superintendents, lawn care operators, and landscapers. It contains the technical information necessary for decision-making in all areas of turfgrass management, as well as practical information useful in the day-to-day operation of turfgrass management programs.

Although the emphasis is on diseases, all aspects of turfgrass culture are touched upon. In addition to helping readers solve the riddle of which grasses are resistant and which are susceptible to various diseases, the book presents cultural and biological programs designed to keep the incidence of disease to a minimum on susceptible grasses, and fungicide recommendations for diseases that require such management. The book presents management strategies that take into account all aspects of turfgrass management (i.e., diseases, cultivation, fertilization, irrigation, mowing).

This book differs from most texts on turfgrass management in that it is not simply a compilation of facts which the reader must evaluate. It lays out specific strategies that can be followed to prevent diseases from occurring. Not everyone can interpret which facts are valid and which are not. Interpretation and management strategies for successful turfgrass management is what this book is all about. It must be remembered that turfgrass management is an art based on science.

I wish to sincerely thank James B. Beard, George W. Bird, Don Blasingame, Herb Cole, Phil Colbaugh, T. E. Freeman, Charles Gould, Bobby G. Joyner, W. A. Meyer, Noel Jackson. George Kozelnicky, Charles Laughlin, Bob E. Partyka, Kenyon T. Payne, Paul Rieke, Pat Sanders, Robert Shearman, Richard Smiley, Ward Stienstra, and A. J. Turgeon for editing various portions of the original text and for their excellent suggestions. Thanks also to Carolyn Brown, Sarah Pounder, and Chris Weiler, who gave valuable technical assistance, as did Hope Shepard, who also helped type the original manuscript, and to Linda Swain, who spent long hours typing and organizing the original manuscript. A special thanks to Julia Roesler for the fine artwork in this book.

This second edition was basically a rewriting of the book because so much has changed in our knowledge of turfgrass science in the 14 years since the writing of the original book. A special thanks goes to Pat Sanders, Gail Schumann, and James B. Beard who reviewed the entire second edition and whose editing and suggestions greatly improved the quality of this book. I also want to thank Gail Conser for typing part of the second edition. A very special thanks goes to Vivian Collier at Lewis Publishers for the long hours she spent editing and organizing the book and to Nancy Dykema for helping me edit and compile the book.

I wish to thank my long-time associate A. Ronald Detweiler, who ran my research program so I could devote the necessary time to write this book.

I am also grateful to Frank Howard, for having faith in me; Harry Young, for taking a chance on me and convincing me to stand up for what I believe in; Roy Wilcoxon, for teaching me to deal harmoniously with my fellow scientists and employees; and Gene Saari, who convinced me that to be anything less in this life than a plant pathologist would be to fail. Others who have inspired me in my career are Mike Prescott, Bob Kroll, Jim Percich, Chuck Laughlin, Bill Fields, Don Benham, and Gordon La Fontaine.

A special thanks to Mary Ann, Jay-Jay, Michael, and Katie for their patience and understanding during the long hours I neglected them while writing both editions of this book.

Contents

Advances in Turfgrass Science

Management of Turfgrass Diseases

SECOND EDITION

CHAPTER 1

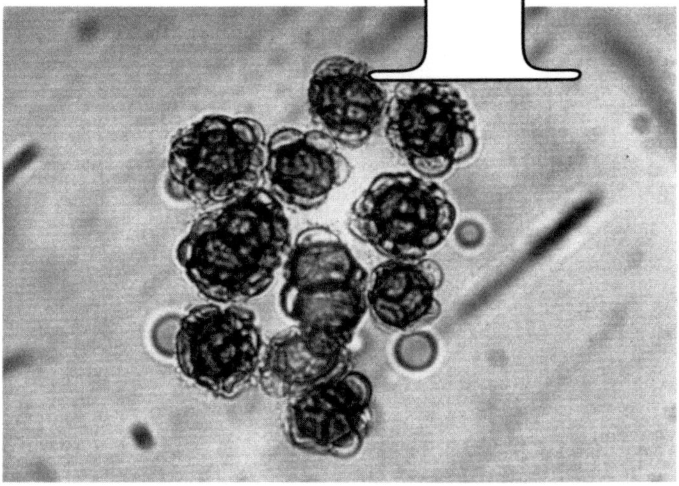

Introduction to Turfgrass Diseases

CAUSES OF TURFGRASS DISEASES

There are five groups of organisms that cause plant diseases—fungi, bacteria, viruses, nematodes, and mycoplasma. Fungi are the most important cause of turfgrass diseases; they are followed in importance by the nematodes and the viruses. There is only one major viral disease of turfgrass, St. Augustine decline, and one major bacterial disease, bacterial wilt.

A disease is an abnormality in structure or function that is caused by an infectious agent and that injures the plant or destroys its aesthetic value. Diseases are sometimes classified into two categories, infectious and noninfectious (or physiological), but I prefer to consider noninfectious diseases as injuries. Injury is damage to a plant that is caused by a noninfectious agent and that injures the plant or destroys its aesthetic value. Damage from hail, lightning, nutrient deficiencies, and fertilizer or pesticide burn are examples of noninfectious plant injury.

A pathogen is an agent that causes a disease. Most pathogens are parasites as well. A parasite is an organism that obtains some or all of its nutrients from a living host. (The organism that a parasite lives on is called its host). Some pathogens are obligate parasites; they can obtain their nutrients only from a live host or living tissue. Rusts, powdery mildews, and all viruses are obligate parasites. Organisms that live only on dead organic matter are called saprophytes (slime molds are an example). Organisms that are mostly parasitic but that can, under certain conditions, live as saprophytes are called facultative saprophytes (for example, *Typhula, Sclerotinia*, and *Helminthosporium*). Organisms that live most of the time as saprophytes but that can, on occasion, become parasites are called facultative parasites (for example, *Rhizoctonia* and *Pythium*). Table 1.1 shows the classification of some organisms that are known to cause diseases in turfgrass, and Figures 1.1 and 1.2 illustrate the developmental stages of the facultative saprophyte *Drechslera* spp.

In order for disease to occur, three conditions are necessary: a susceptible host, a virulent pathogen, and a favorable environment. These three conditions are called the plant disease triangle. If any one of them is missing, disease will not develop.

The disease process usually involves four steps. They are infection, incubation, symptom development, and inoculum production. Infection is the process by which a disease-producing organism (pathogen) enters the plant. Incubation is the period during which the pathogen inhabits its host without producing visible symptoms. The interaction between the pathogen and its host results in symptom development. Inoculum production is the process whereby the pathogen reproduces propagules for spread and survival. The inoculum can be simply the

Table 1.1 A Classification of Organisms That Cause Turfgrass Diseases Based on Life Cycles

Obligate Parasites	Facultative Saprophytes	Facultative Parasites	Saprophytes
Erysiphe graminis	*Bipolaris sorokiniana*	*Colletotrichum graminicola*	Fairy rings*
Parasitic nematodes	*Drechslera poae*	*Gaeumannomyces graminis*	*Mucilago crustacea*
Puccinia spp.	*Gloeocercospora sorghi*	*Leptosphaeria korrae*	*Physarum cinereum*
St. Augustine decline virus	*Laetisaria fuciformis*	*Pythium aphanidermatum*	
	Limonomyces roseipellis	*Rhizoctonia solani*	
	Microdochium nivale	*Xanthomonas campestris*	
	Pyricularia grisea		
	Sclerophthora macrospora		
	Sclerotinia homoeocarpa		
	Typhula spp.		
	Ustilago striiformis		

*Mild infection may occur.

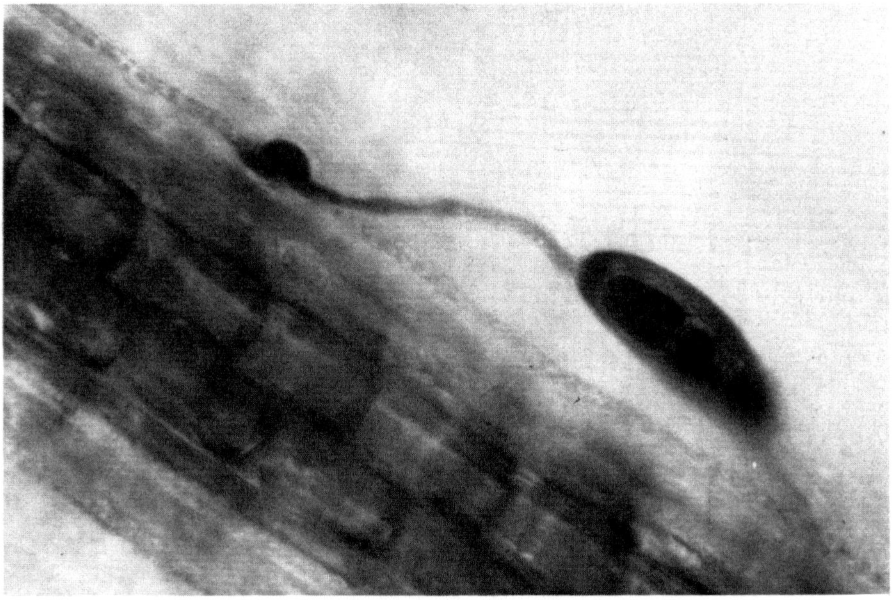

Figure 1.1. A *Drechslera poae* spore on a fescue leaf with germ tube that has produced an appressorium. A penetration peg will be produced by the appressorium which will infect the leaf.

spore of a fungus, or it can be the entire organism, as it is in the case of a virus or a bacterium.

IMPORTANCE OF TURFGRASS DISEASES

Disease plays a major role in determining the success or failure of a turfgrass stand. It is often the most important single factor limiting the successful growth of a cultivar or species, a fact that you must keep in mind when selecting a turfgrass species or cultivar. For example, spring dead spot (SDS) on bermudagrass limits its widespread use as a fairway grass, especially in the northern regions of the warm-season grass belt; St. Augustine decline (SAD), a viral disease, has eliminated St. Augustinegrass as a desirable turfgrass in many areas; and necrotic ring spot, summer patch, and stripe smut have made it impractical to plant Kentucky bluegrass cultivars like Merion, Windsor, and Fylking. The best-textured, nicest-colored, and fastest-germinating grass, if it is susceptible to a major pathogen, will turn into the worst-looking and poorest-colored grass imaginable when it becomes decimated by a disease, and you will wish it hadn't germinated at all.

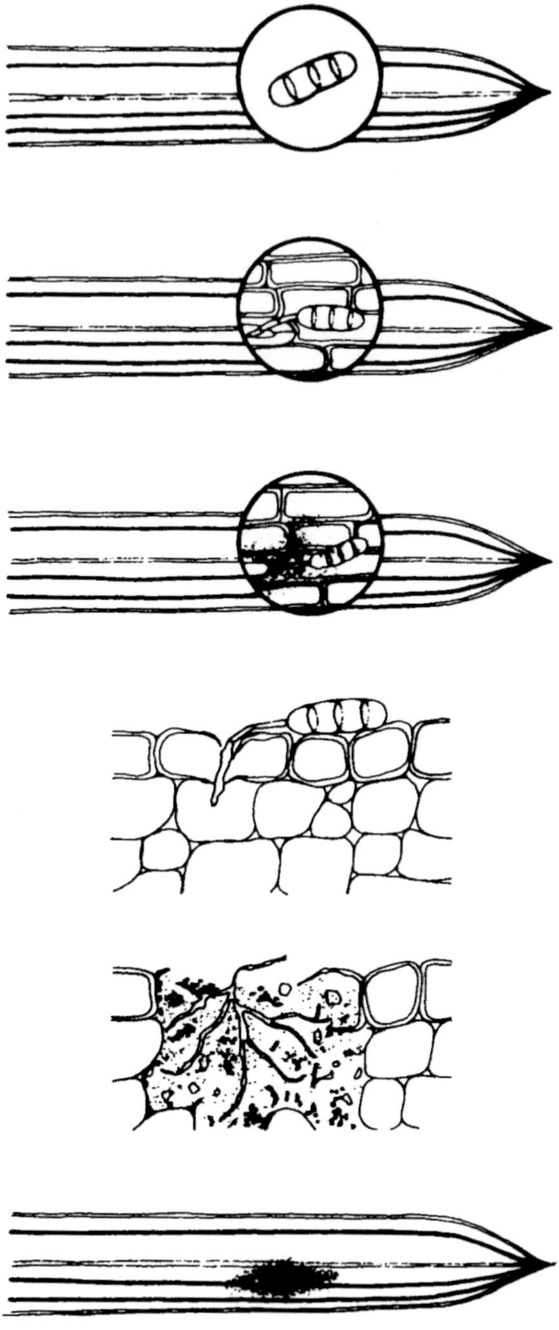

Figure 1.2. Germination of a *Drechslera* spore and subsequent penetration, incubation, and development of symptoms on a turfgrass blade.

Turfgrass diseases, caused by fungi, are of great economic importance. It is difficult to get exact figures, but the turfgrass industry spent 80 million dollars on fungicides in 1988. More fungicides are used on turfgrass than on any other single crop in the United States.

IDENTIFYING TURFGRASS DISEASES

A little knowledge is said to be a dangerous thing—and it is! Often samples of dead or dying grass are collected and mailed, with little background information, to a diagnostic laboratory. At the laboratory, disease organisms are isolated from the plant material and a diagnosis is made. Under such circumstances a disease may be diagnosed as "Helminth" or as necrotic ring spot, when it is neither. Or what appear to be black *Helminthosporium* lesions are observed on dying turf, and it is assumed that *Helminthosporium* is causing the problem, which may or may not be the case. It would be difficult to find a patch of Kentucky bluegrass in the spring that didn't have *Helminthosporium* lesions, but does that mean this organism is causing the problem? Maybe. Before you jump to conclusions, look at the other symptoms in the area. Ask such questions as: Were there many black lesions? Were they large? Did the grass show a general, all-over thinning? If the answers are yes, then *Helminthosporium* is probably the cause. Were "frog-eye" patterns present? Was there evidence of wilt inside the "frog-eyes? Did hillsides and southern exposures get the disease first and most severely? When wilting occurred, did syringing seem to help? If the answer to all of these questions is yes, then it is likely that necrotic ring spot is the problem.

Koch's Postulates

Koch's postulates provide a basis for establishing the cause of a disease.

1. **Association**—The organism suspected of causing the disease must always be present when the disease occurs.
2. **Isolation**—It must be possible to isolate the suspect organism and grow it in a pure culture (except in the case of obligate parasites).
3. **Inoculation**—When a disease-free host is inoculated with the organism, it must exhibit the same symptoms observed when the disease occurs spontaneously.
4. **Reisolation and comparison**—When the organism suspected of causing the disease is reisolated from the inoculated plants, grown in a

pure culture, and then compared to the first organism isolated, the two organisms must be identical.

It is important to satisfy all of Koch's postulates when identifying the causal organism of a disease. A good example of what can happen when all the steps of Koch's postulates are not followed is the case of *Fusarium* blight. The causal organisms of this disease were determined to be *Fusarium roseum* and *F. tricinctum* based on isolations from the field and greenhouse inoculations that produced foliar lesions on Kentucky bluegrass.[24] The researchers[24] failed to demonstrate "frog-eye" symptoms, and therefore never completed steps 3 and 4 of Koch's postulates. This led to many years of confusion in understanding the epidemiology and management of this disease. It wasn't until many years later when researchers began to reexamine the disease that the true causal agents were found, *Magnaporthe poae* and *Leptosphaeria korrae*, and the disease was renamed from *Fusarium* blight to summer patch and necrotic ring spot, respectively. Yes, these fungi did produce "frog-eye" patterns in both the laboratory and field when the correct pathogens were used for inoculations.

Taking Samples

When you take samples of diseased grass, follow these instructions:

1. Look for an area that has just begun to show symptoms. Older areas may be contaminated with saprophytic organisms living on the dead tissue.
2. Take a large enough sample. You need at least a 6-inch plug, not just a couple of leaves. Include the root system of the plant in your sample.
3. If the sample is to be mailed, wrap it in paper. Never put it in a plastic bag, because the saprophytic fungi, bacteria, and nematodes will turn the sample to mush before it arrives. The one exception is soil samples for nematode analysis, which should be placed in a plastic bag to keep them from drying out. Samples should be packed tightly and mailed immediately after they are taken and early in the week, so that they don't sit in the post office over the weekend.

Include the following information with your samples:

1. species and cultivar of turfgrass
2. age of the grass stand
3. symptoms (wilt, yellowing, stunting, leaf spots, thinning, etc.)
   a. initial symptoms

   b. current symptoms
   c. pattern of symptoms on the grass stand
4. prevalence of the disease (a few spots, large areas, all greens, one green, etc.)
5. severity of the disease (moderate, severe)
6. location of the disease (high spots, low spots, edges of greens, near house, under tree, etc.)
7. soil type (clay, sand, loam, heavy soil, etc.)
8. recent treatments
   a. fertilizer (date of last application, amount and type)
   b. pesticides (not just fungicides) applied in last month
   c. cultivation (give dates)
      (1) coring (aerifying)
      (2) spiking
      (3) vertical mowing
9. weather
   a. just before symptoms developed
   b. at the time symptoms developed
10. date the problem was first observed

Interpreting Research

It is often difficult for laypersons to interpret the data supplied by researchers. The researcher may tell only half the story or may not relate laboratory findings to what actually happens in the field. For example, Fylking is still reported to be resistant to melting-out, which it is, and which sounds good since melting-out is a major disease. But if the researcher fails to mention that Fylking is susceptible to necrotic ring spot (see Figure 1.3) and summer patch—as a matter of fact, so susceptible that either of these diseases has the potential to completely destroy the Fylking turf in a matter of a few years—you will be surprised and disappointed when your stand of grass becomes decimated, yet I have seen this happen. Sometimes, after expounding a variety's good points for 10 minutes, a scientist will end by saying "Of course it has some other problems," or "It is, of course, susceptible to summer patch." Unless you know how susceptible it is to summer patch or how serious a disease summer patch is, you will probably plant it, thinking, "Well, it only has one problem compared to so many good points." It would be like buying a car that was perfect except for a defective motor. In both cases you waste a lot of money on something that won't work.

The same thing happens with data on disease. Researchers have published results and given talks stating that Kentucky bluegrass is more susceptible to rust, "Helminth," or red thread at high nitrogen levels

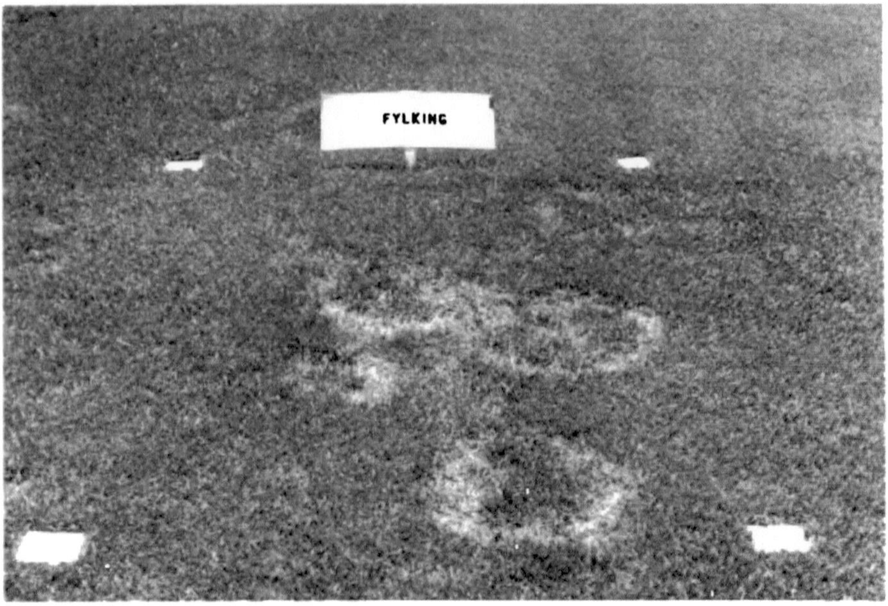

Figure 1.3. Necrotic ring spot in Fylking Kentucky bluegrass.

than at low levels. It is true, too—in the greenhouse or laboratory, where grass is grown in pots, fertilized, and artificially inoculated, and where readings are taken before the grass is mowed or clipped. But in the field, nitrogen can help control "Helminth" rust and red thread on Kentucky bluegrass. Why? The difference is that in the field the grass is mowed at least once a week. Since rust, "Helminth," and red thread take from 10 to 14 days to develop after the grass is infected, and all are foliar pathogens, weekly mowing will ensure that the infection is mowed off before the disease develops. Such laboratory data do increase our understanding of the disease, but they belong in scientific journals and at scientific meetings, not in journals read by lay people and not at turfgrass conferences and other grower meetings. Why explain laboratory results to a person whose livelihood is growing grass in the field, if what happens in the laboratory has no correlation with what happens in the field? Be wary of laboratory data that are not accompanied by good field data.

This brings up another point. Each year new ideas come forth; some with good data to back them up, others with no supporting data or one year's experimental results, or merely some random observations. Before you rush headlong into trying a new product or cultural practice that may cost you your job, ask to see the data. If you have trouble understanding the data, ask your local turf expert for an evaluation. If a cultural practice or chemical control could result in the loss of turf, try it on a small area first to see how it works under your conditions, regardless of the advice you have been given. Try thinking for yourself!

DISEASE MANAGEMENT

The word *management* is used in this book in preference to the word *control*. Control implies the eradication of a disease problem. In reality, however, the problem will occur season after season or recur many times during a single growing season. Therefore, diseases actually are managed rather than controlled, and an approach that stresses management should prove more realistic and useful to the turfgrass manager than one that assumes complete control is possible.

Too often turfgrass pathologists have dealt with single diseases as though they occur in a vacuum. Too often chemical and cultural recommendations or cultivar selection are based on consideration of only one disease. A single species or cultivar may be attacked by many diseases throughout the growing season, and therefore, the total picture must be evaluated. Recommendations of cultivars, cultural practices, and chemical agents must reflect a knowledge of all the diseases that can occur on a given species or cultivar throughout the season.

Plant pathologists also need to stop talking about diseases as though they occur on just any green vegetation. Diseases attack specific grass species, or, in some cases, specific cultivars. Turfgrass diseases should be thought of in relation to the specific cultivar or species they afflict.

Strategies for dealing with turfgrass diseases include planting cultivars that are resistant to the major diseases, employing appropriate cultural practices as well as fungicide programs. Chapters 11 and 12 discuss such strategies.

Table of Contents for Color Plates

PLATE 1
Anthracnose attacking an untreated annual bluegrass fairway with healthy annual bluegrass treated green and approach.

PLATE 2
Anthracnose damage on an annual bluegrass-creeping bentgrass green; the annual bluegrass has been severely thinned and yellowed, while the creeping bentgrass remains unaffected.

PLATE 3
Acervuli and setae of *Colletotrichum graminicola* occurring in rows on infected annual bluegrass blade.

PLATE 4
Injury from the application of granular sulfur to a creeping bentgrass green. Spots often resemble dollar spot symptoms.

PLATE 5
Close-up of dollar spot on a Toronto creeping bentgrass.

PLATE 6
Dollar spot with fungal mycelium present. Notice cobweb of small spider evident in upper left-hand corner which is often confused with dollar spot mycelium. Notice no dead turf under cobweb.

PLATE 7
Brown patch in St. Augustinegrass (Photograph courtesy of Dr. Don Blasingame).

PLATE 8
Brown patch on St. Augustinegrass lawn.

PLATE 9
Brown patch on zoysiagrass turf.

PLATE 10
Brown patch on creeping bentgrass green with "smoke ring" evident.

PLATE 11
Brown patch on annual bluegrass green.

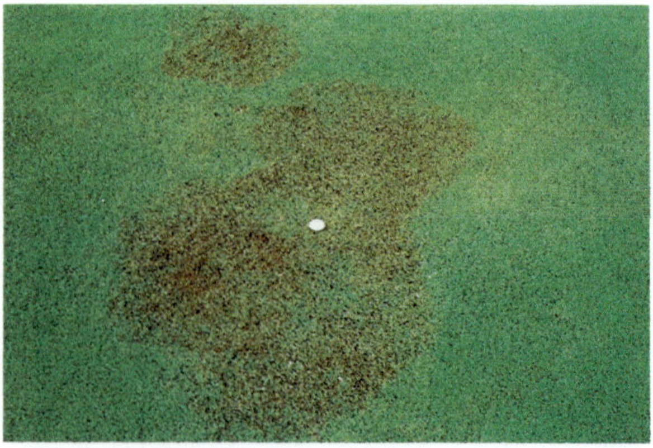

PLATE 12
Brown patch on creeping bentgrass green.

PLATE 13
Pythium blight on perennial ryegrass fairway with streaking on banks. Notice the strip of crabgrass on the edge of fairway that is unaffected by the disease.

PLATE 14
Copper-colored spots of *Pythium* blight on a creeping bentgrass fairway.

PLATE 15
Pythium blight streaking on a perennial ryegrass fairway.

PLATE 16
Summer patch on Kentucky bluegrass; notice the irregularly shaped wilting patches.

PLATE 17
Close-up of summer patch on annual bluegrass fairway.

PLATE 18
Summer patch on annual bluegrass fairway.

PLATE 19
Summer patch eliminating the annual bluegrass from the fairway and leaving the creeping bentgrass unaffected.

PLATE 20
Summer patch eliminating the annual bluegrass from an annual bluegrass-creeping bentgrass green, leaving the creeping bentgrass unaffected.

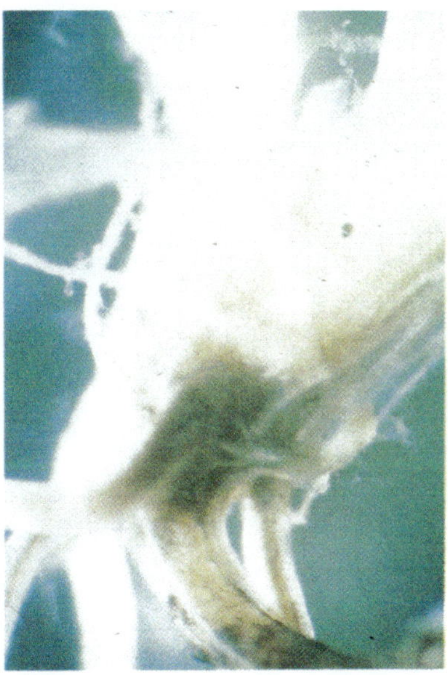

PLATE 21
Necrotic crown on an annual bluegrass plant infected with *Magnaporthe poae*. Runner hyphae on root, along with mycelial mats.

PLATE 22
Several patches of bermudagrass decline on a bermudagrass green.

PLATE 23
Bermudagrass decline on a bermudagrass green.

PLATE 24
Spring dead spot in a bermudagrass turf (Photograph courtesy of Dr. Don Blasingame).

PLATE 25
Typical "frog-eye" symptoms of necrotic ring spot.

PLATE 26
Active necrotic ring spot patch with typical "frog-eyes" pattern. Photograph taken in November; notice red leaf blades in outer margin.

PLATE 27
Close-up of red Kentucky bluegrass blades in a necrotic ring spot patch. The red leaf blades are common when the disease is active during the cool weather of the fall.

PLATE 28
Summer symptoms of necrotic ring spot. Notice straw-colored appearance. Also notice how much more disease is evident in the lower height of cut on the fairway than the taller cut rough.

PLATE 29
A healthy plot of Midnight Kentucky bluegrass surrounded by cultivars of Kentucky bluegrass that are susceptible to necrotic ring spot.

PLATE 30
Take-all patch on creeping bentgrass green with unaffected annual bluegrass invading the center of the patch.

PLATE 31
Pink patch on a colonial bentgrass turf.

PLATE 32
Pink patch on Pennlinks creeping bentgrass green.

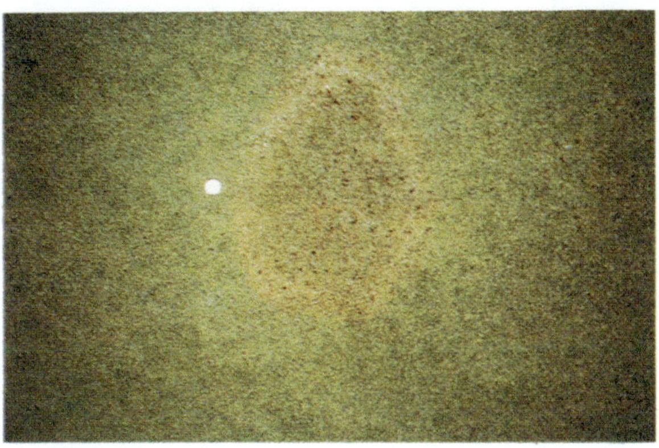

PLATE 33
Patch of unknown etiology on a *Paspalum vaginatum* green in South Africa.

PLATE 34
Crown rotting anthracnose on a green.

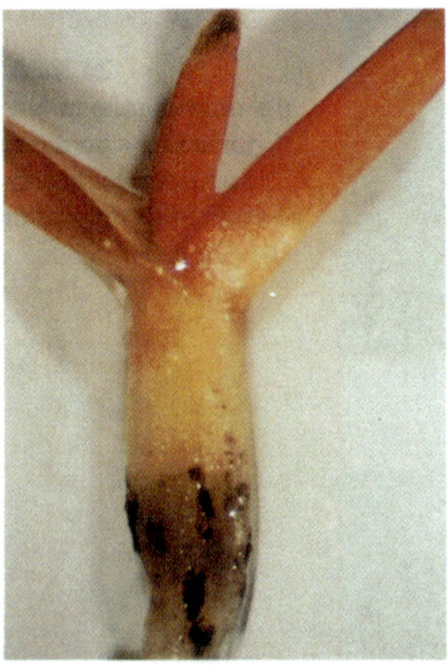

PLATE 35
Crown rotting anthracnose infected annual bluegrass plant; notice acervuli in crown region.

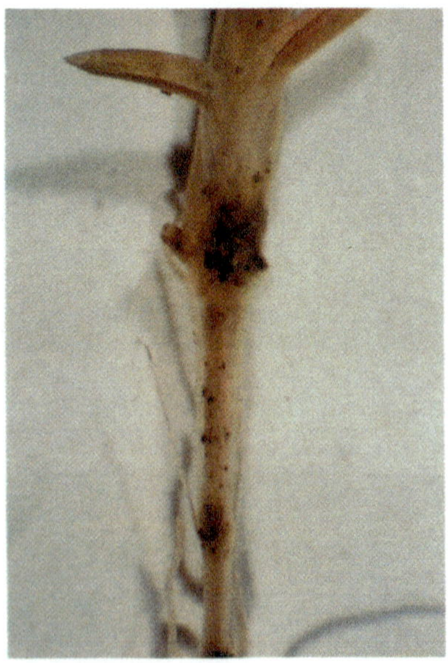

PLATE 36
Crown rotting anthracnose on crown of intact plant.

PLATE 37
Microdochium patch on a Penncross creeping bentgrass turf during the fall.

PLATE 38
Microdochium patch (pink snow mold) on a Penncross creeping bentgrass turf in the spring after the snow has melted.

PLATE 39
Microdochium patch on annual bluegrass green during the spring of the year.

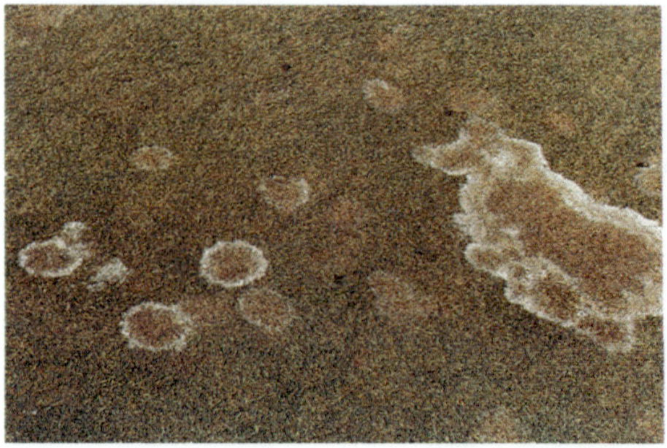

PLATE 40
Typhula blight (gray snow mold) showing the grayish colored mycelium of the fungus in a Penncross creeping bentgrass turf.

PLATE 41
Sclerotia of *Typhula incarnata* in Penncross creeping bentgrass immediately after the snow has melted.

PLATE 42
Basidiocarps of *Typhula incarnata* which arose in the fall from sclerotia that were produced the previous spring and over-summered in the thatch.

PLATE 43
Typhula blight on creeping bentgrass turf. Dark patches caused by *Typhula incarnata*, lighter-colored patches caused by *Typhula ishikariensis*.

PLATE 44
Close-up of mycelium of *Typhula* blight.

PLATE 45
Pythium blight in overseeded ryegrass (Photograph courtesy of Dr. Don Blasingame).

PLATE 46
A stripe smut infected Kentucky bluegrass plant.

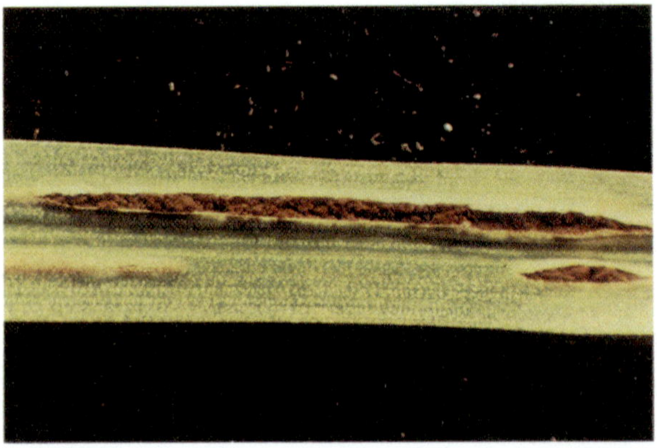

PLATE 47
Close-up of stripe smut on Merion Kentucky bluegrass blade; note the brown teliospores erupting through the epidermis.

PLATE 48
Superficial fairy rings in a Kentucky bluegrass turf.

PLATE 49
Rust-colored urediospores of bermudagrass rust on a Sunturf bermudagrass leaf.

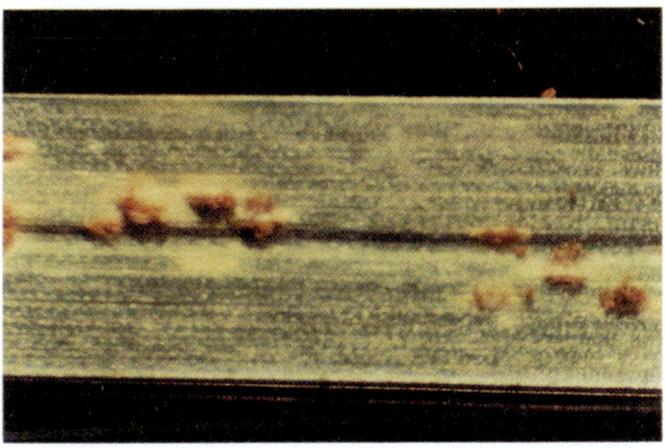

PLATE 50
Close-up of rust pustules on Merion Kentucky bluegrass.

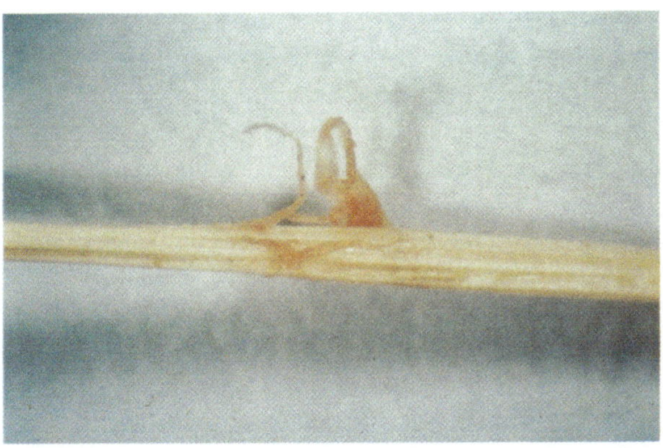

PLATE 51
Red thread stroma of *Laetisaria fuciformis* on a perennial ryegrass leaf blade.

PLATE 52
Red thread in a perennial ryegrass lawn.

PLATE 53
Laetisaria red thread in a fine-leaf fescue turf.

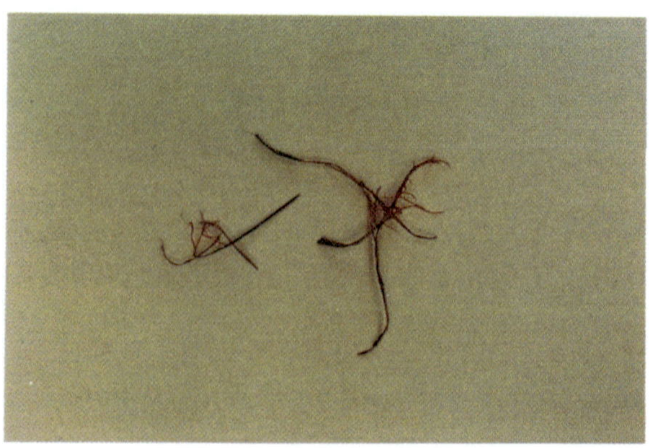

PLATE 54
Close-up of red thread on fine-leaf fescue.

PLATE 55
A plot of Palouse Kentucky bluegrass being destroyed by melting-out compared to a healthy plot of Nuggett Kentucky bluegrass which is resistant to the disease.

PLATE 56
Melting-out on Kentucky bluegrass; notice the large lesion on both the leaf blades and sheath.

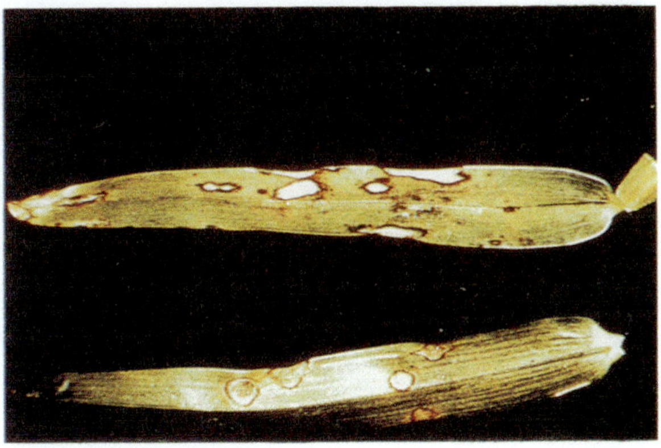

PLATE 57
Foliar symptoms of gray leaf spot on St. Augustinegrass (Photograph courtesy of Dr. Don Blasingame).

PLATE 58
Yellow tuft infected turfgrass plants; the two on the left are creeping bentgrass, while the two on the right are Kentucky bluegrass. Notice the poor root systems on the plants.

PLATE 59
Yellow tuft in a Penncross creeping bentgrass turf.

PLATE 60
Close-up of St. Augustine decline symptoms on turfgrass blades (Photograph courtesy of Dr. Bobby G. Joyner).

PLATE 61
Powdery mildew on blades of Merion Kentucky bluegrass.

PLATE 62
Slime mold fungus living as an epiphyte on turf.

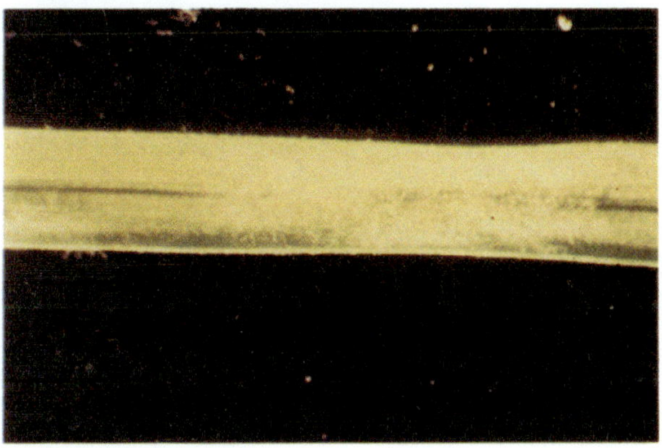

PLATE 63
Close-up of powdery mildew on Merion Kentucky bluegrass blade.

PLATE 64
Stroma of red thread fungus.

PLATE 65
Close-up of anthracnose acervuli on a leaf of annual bluegrass.

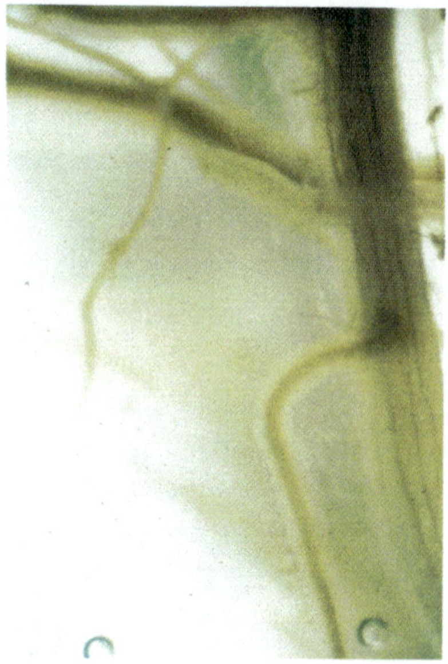

PLATE 66
Ectotrophic hyphae (runner hyphae) of *Magnaporthe poae* on an annual bluegrass root.

PLATE 67
A Toronto creeping bentgrass green destroyed by bacterial wilt.

PLATE 68
Close-up of bacterial wilt destroying individual plants on a Toronto creeping bentgrass green.

PLATE 69
Close-up of phosphorus deficiency in a high sand content green resulting from not adding phosphorus in fertilizer. Often confused with disease symptoms.

PLATE 70
Turf loss on golf green from effects of black layer.

PLATE 71
Black layer in creeping bentgrass green from sulfur applications.

PLATE 72
Black layer produced in sand. Notice the blackness occurs only in layer where sulfur was added.

CHAPTER 2

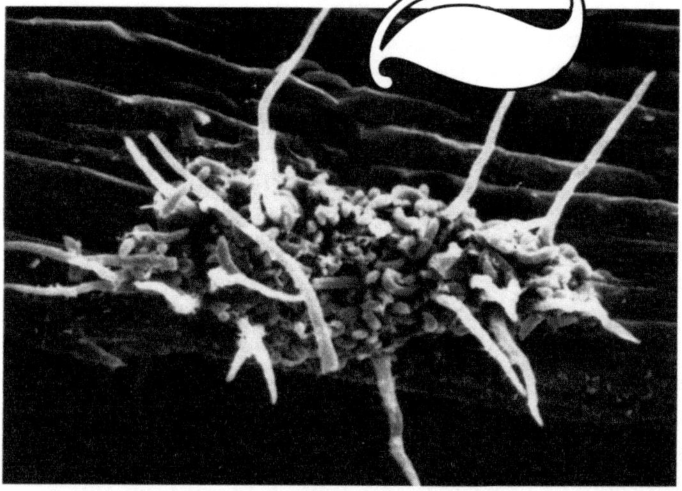

Fungal Diseases of Turfgrass I: Diseases Primarily Occurring on Golf Course Turfs

DISEASE: Anthracnose

Pathogen: *Colletotrichum graminicola* (Ces.) Wils. (syn. *C. cereale* Manns)

Hosts: Annual bluegrass, creeping bentgrass, Kentucky bluegrass, fine-leaf fescue, perennial ryegrass bermudagrass

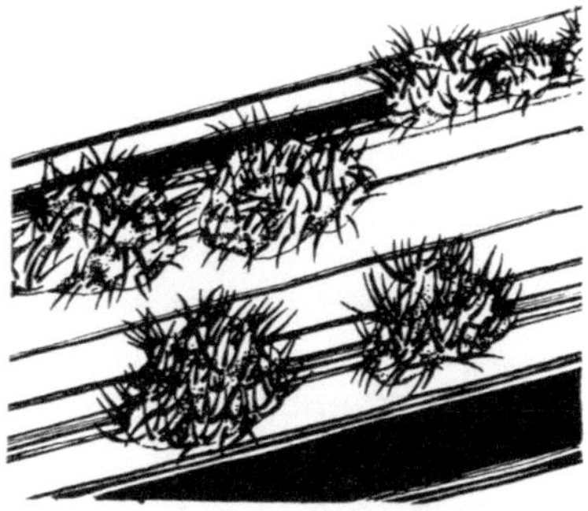

Colletotrichum graminicola acervuli with setae

Anthracnose has been described as a minor disease of annual bluegrass and several other turfgrasses in England.[133] In the United States it has been reported to be one of the key factors causing death of annual bluegrass during warm weather.[1,154,157] It appears to be one of the factors that limit the successful growth of fine-leaf fescues in the warmer areas of the cool-season grass belt, and Duell and Schmit have reported that anthracnose eliminates perennial ryegrass in roadside turfs.[42]

Before 1975, the dying of *Poa annua* (annual bluegrass) in summer was attributed to "wilting" (direct high-temperature kill) or the natural dying of a winter annual. Many reports[31,157,158] showed that fungicide treatments would prevent annual bluegrass from dying during summer stress periods. They showed that annual bluegrass was not dying from heat stress, drought stress, or the natural dying of a winter annual, but from other causes that could be mitigated by fungicides. The summer die-off was all originally attributed to the fungus *Colletotrichum graminicola*, the causal agent of the disease anthracnose. Subsequent research disclosed that other factors may be involved in the summer decline of annual bluegrass, including summer patch and the black turfgrass ataenius beetle. Through the use of fungicides, insecticides, and good cultural practices described in Chapter 11, annual bluegrass can be successfully grown in most cool-season grass-growing areas of the world.

Symptoms

Anthracnose is characterized by irregularly shaped patches of yellow-bronze turf, ranging in size from a few inches to several feet. Leaf lesions initially appear as elongated reddish brown spots on the leaves. These spots may enlarge, eventually encompassing the entire leaf blade. Numerous black fruiting bodies (acervuli) of the fungus can be seen on the foliage as the disease progresses. The acervuli have black spines (setae) protruding from them.

Occurrence

C. graminicola can overwinter as a saprophyte in plant debris or as a pathogen in infected tissue. Its life cycle is illustrated in Figure 2.1. It appears to attack turf growing under conditions of low soil fertility, high temperature stress, or both. It can show up on the roots, stolons, or leaves, but is most commonly found on the foliage of the grass plant. *C. graminicola* usually infects grass blades from the tip down, especially grass that has been freshly mowed. The serious damage to the turf occurs during periods of high temperature and humidity when the

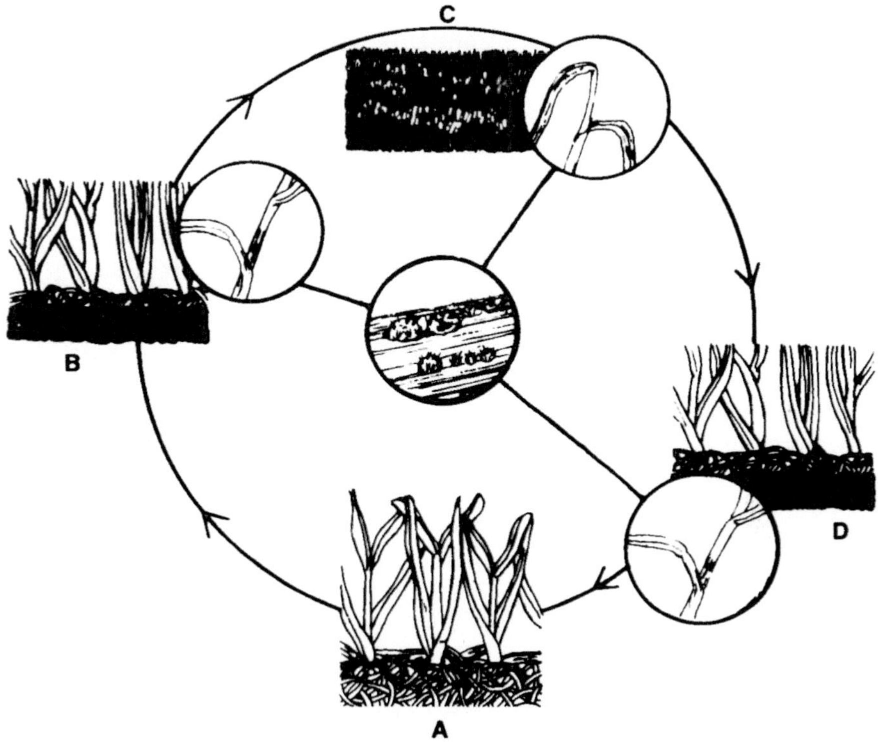

Figure 2.1. Life cycle of *Colletotrichum graminicola*. (A) The fungus overwinters in dead plant tissue in the thatch and on stems beneath the leaf sheath; (B) Visual infections in the spring are confined mostly to the stems beneath the leaf sheaths. The main symptom is yellowing of the plant. (C) When the weather becomes hot, the plants turn bronze and die. Acervuli are evident on the leaves. (D) Acervuli disappear from leaf blades as cool weather returns, but are present on stems beneath the leaf sheath. Yellowing of the plant occurs again. (E) Close-up of acervuli on an infected plant.

turf is under severe stress. The disease is most severe in areas of compaction, heavy traffic, and poor soil drainage.

MANAGEMENT

Cultural

Anthracnose can be managed with light nitrogen application (one-half pound per 1000 square feet [.25 kg per 100 square meters] every three weeks) when it occurs during cool weather. The same light nitrogen applications will help to reduce the severity of the disease during warm weather, but fungicide treatments will also be necessary to prevent severe turf loss.

Genetic

Creeping bentgrass appears to be resistant to anthracnose in the northern region of the cool-season grassbelt, but it is reported to be susceptible in the southern regions, the Kentucky bluegrass-bermudagrass transition zone, and the warm-season grassbelt. Even in the northern region of the cool-season grassbelt, a wide range of resistance and susceptibility probably exists among creeping bentgrass cultivars. Furthermore, although Kentucky bluegrass, perennial ryegrass, and the fine-leaf fescues are listed as susceptible, there is undoubtedly a great range of resistance and susceptibility among the many cultivars of these grasses.

Chemical

Anthracnose can be managed with systemic fungicides like benomyl, fenarimol, propiconazole, thiophanate methyl, and triadimefon applied every 2 to 3 weeks, or with contact fungicides like chlorothalonil, mancozeb, or maneb plus zinc sulfate applied every 7 to 10 days.

DISEASE: Crown Rotting Anthracnose (Basal Rot Anthracnose)

Pathogen: *Colletotrichum* spp.

Hosts: Annual bluegrass, creeping bentgrass

Crown rotting anthracnose attacks the crowns of grass plants, unlike the more common anthracnose which primarily attacks the foliage. This is a devastating disease in which infected plants normally die.

Symptoms

Crown rotting anthracnose appears as bronze-yellow spots in the turf, usually about the size of a dime. Examination of infected plants reveals darkened crowns which appear as though someone had stuffed charcoal in them. When viewed under magnification, the acervuli and setae of the fungus can be seen.

Occurrence

The disease occurs primarily on annual bluegrass throughout most of the world. However, it has been reported to be a widespread serious problem on creeping bentgrass in the eastern United States. It is most noticeable on greens-height mowed turf. It appears to be more common on sand or sand-top dressed greens. Very little is known about the epidemiology of the disease. It was originally called cold temperature anthracnose because it was primarily seen in the cool weather of spring. It has since been observed occurring throughout the growing season.

MANAGEMENT

Cultural

No information available.

Genetic

The only creeping bentgrass it has been reported on is "Penncross," but this may be because it has been so widely grown. Only time will tell if the newer cultivars are truly resistant. My guess is they are not.

Chemical

No curative programs have been developed, probably because so little is known about its development. Greens with multiple applications of the DMI fungicides for summer patch management still develop the disease, indicating they are either not effective or the timing is wrong. Curative application of the DMI fungicides has not been effective.

The benzimidazole fungicides applied through drench application at patch disease rates have been the most effective treatment in stopping the spread of the disease. Infected plants die, regardless of treatment.

DISEASE: Brown patch (*Rhizoctonia* blight)

Pathogen: *Rhizoctonia solani* Kühn

Hosts: Annual bluegrass, bahiagrass, bermudagrass, centipedegrass, colonial bentgrass, creeping bentgrass, fine-leaf fescues, Kentucky bluegrass, meadow fescue, perennial ryegrass, St. Augustinegrass, tall fescue, velvet bentgrass, zoysiagrass

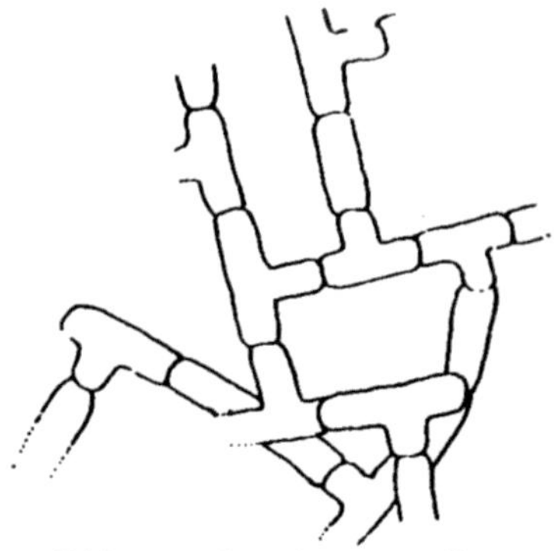

Rhizoctonia solani mycelium

Brown patch is the most widespread of all the turfgrass diseases. It occurs in all areas where turf is grown, and it attacks all known turfgrasses. *Rhizoctonia solani* has one of the widest host ranges of agricultural crops of any plant pathogen on agricultural crops, in part because it is a soil inhabitor; that is, it lives on organic matter present in the soil and competes well with the other saprophytic microorganisms.

Symptoms

The disease occurs in the turf as circular patches, from a few inches up to several feet (few cm to a couple of meters) in diameter. The infected leaves first appear water-soaked and dark, eventually drying, withering, and turning dark brown. Brown to black sclerotia are sometimes found beneath the leaf sheath or on the crowns and roots. When the humidity is high a "smoke ring," which consists of the mycelium of the fungus, surrounds the outer margins of the diseased area (Figure 2.2). The "smoke ring" will disappear as the foliage dries. It is *not* necessary to have a "smoke ring" in order to have brown patch. Too often superintendents say, "I have what looks like brown patch but it can't be, because there is no smoke ring." While the "smoke ring" aids in identifying brown patch, it is not always present, and only appears when the turf is wet and the humidity is high.

In the early stages of brown patch, only a few leaves and grass plants are destroyed, but if the disease is not checked immediately many more grass plants will be lost, especially on low-cut turf such as golf course greens. If warm weather continues and treatment is not applied, algae can cover the turf area infected by brown patch, eventually forming a hard crust which makes recovery even more difficult.

Another identifying characteristic of brown patch is a musky odor

Figure 2.2. ***Rhizoctonia*** **brown patch with "smoke ring" evident on a golf course green (Photograph courtesy of Dr. Bill Daniels).**

perceptible 12 to 24 hours before the disease appears and sometimes persisting after the disease appears. Some old-time superintendents spray "as needed," using the musky smell as an indicator of when to spray.

The disease takes quite a different form in California. Madison[99] reports brown patch as being primarily a disease of the roots and crowns. The damage to the grass is in the form of a ring, with the grass in the center remaining fairly healthy. The outer ring of grass turns brown and dies, usually from desiccation due to its damaged root system.

Occurrence

R. solani survives adverse periods as sclerotia or as mycelium in plant debris. It can also survive as a saprophyte in the thatch. When the soil temperature rises into the 60s (15–20°C), the sclerotia begin to germinate and the fungus begins to grow. The fungus grows in a circular pattern, as most fungi do, but apparently does not parasitize the grass foliage until the air temperature rises into the mid-80s (approximately 30°C) with high humidity and nighttime temperatures in the 70s

(21–26°C) or higher. *R. solani* is basically a weak parasite. At low temperatures it grows as a saprophyte or causes minute infections that do not seriously damage grass plants that are in good growing condition. It is only after a grass plant suffers heat stress and high-temperature growth stoppage begins that the balance switches to favor the fungus and disease development. One thing is certain: the fungus doesn't grow 6 inches or 3 feet (15 cm or 1 meter) overnight, as some plant pathologists suggest. *R. solani* simply cannot grow that fast. If it did, the entire area would be covered with mycelium instead of just having a "smoke ring" around the outside.

In laboratory experiments, Rowell found the cut leaf tip to be the primary site of entrance of *R. solani*.[124] However, the fungus can also enter through the stomates or by direct penetration of the grass blade. Hearn[68] reported that with the arrival of warm weather the fungus infects roots first, then stolons, and finally leaves.

MANAGEMENT

Cultural

High levels of nitrogen in the field, during or just prior to the onset of hot humid weather, will increase the severity of the disease; therefore, minimum amounts of nitrogen, no more than one-half pound per 1000 square feet (.25 kg per 100 square meters) (actual), and preferably no nitrogen, should be used as hot humid weather approaches. Normal levels of phosphorus and potassium should always be maintained. As with dollar spot, removal of the guttation water (dew) as early as possible in the morning will help to reduce the severity of the disease.

Genetic

There appear to be no resistant species or cultivars. The most susceptible species are tall fescue, perennial ryegrass, St. Augustinegrass, annual bluegrass, and creeping bentgrass.

Chemical

There are contact and systemic fungicides for managing brown patch; however, to be effective they must be used! Brown patch is a disease in which infection takes place long before symptoms are evident, and when the symptoms do occur, large areas of turf are involved. If the disease is not treated right away, and if the weather stays hot and humid, a considerable amount of grass may be lost and recovery may not take place until cool weather returns. Those who rely on a curative program and spray when they smell the musky odor should pray that they don't develop hay fever. By and large, a preventive fungicide program is the best approach.

The DMI fungicides, like propiconazole, fenarimol, and triadimefon, have not worked that well in controlling brown patch. In some instances they have actually increased it. It appears that there are other species of *Rhizoctonia*, specifically *R. maydis* and *R. zeae*, or other strains of *R. solani* that are not controlled by the DMI fungicides. Whatever the case may be, trying to manage brown patch with the DMI fungicides could be dangerous to your current employment status.

DISEASE: Dollar spot

Pathogen: *Sclerotinia homoeocarpa* F.T. Bennett

Dollar spot symptoms on a grass blade

Hosts: Annual bluegrass, bahiagrass, bermudagrass, centipedegrass, fine-leaf fescue, colonial bentgrass, creeping bentgrass, perennial ryegrass, Kentucky bluegrass, creeping and spreading fescue, St. Augustinegrass, velvet bentgrass, zoysiagrass.

Most golf course superintendents consider *Pythium* blight the most important turfgrass disease, yet they spend more money trying to manage dollar spot. This is true throughout most of the temperate and hot humid regions of the United States. On golf courses across the nation more money is spent to manage dollar spot than on any other turfgrass disease.

Symptoms

Dollar spot is found throughout the United States, except in the arid regions of the West. The disease is characterized by round, bleached-out or straw-colored spots ranging from the size of a quarter to the size of a silver dollar (hence the name dollar spot). The spots appear as sunken areas in the turf, especially in turf mowed to half an inch (.15 cm) or shorter. Individual spots may coalesce and destroy the turf in large undefined areas (Figure 2.3). If the fresh spots are observed in the morning while the grass is still wet, the grayish-white, fluffy mycelium

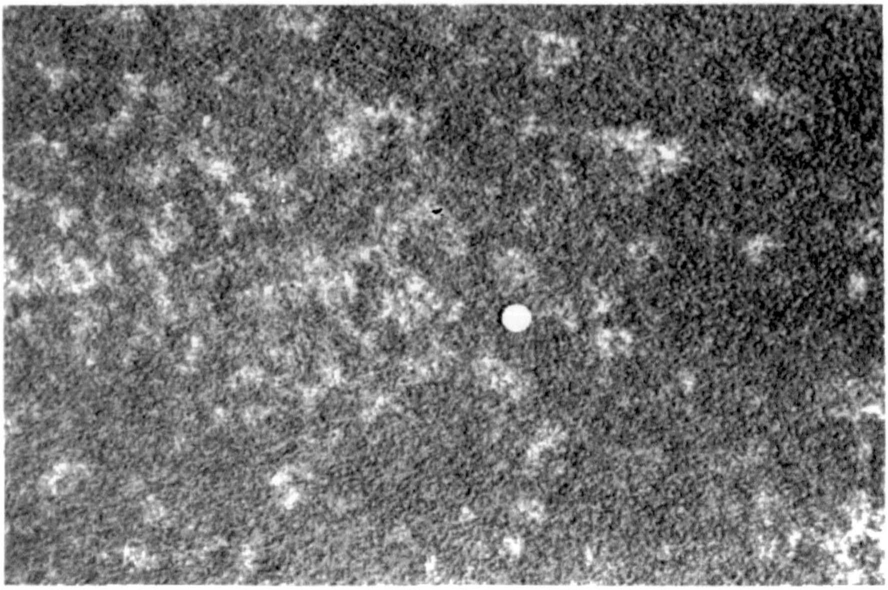

Figure 2.3. Dollar spot on bermudagrass (Photograph courtesy of Dr. Don Blasingame).

of the fungus can be seen. The disease is spread primarily by mowers and other maintenance equipment carrying the mycelium and infected plant tissue. Infected plant tissue is also carried by golf shoes and carts. When the disease is controlled only on the greens, its recurrence will usually be noticed first on the side of the green that gets traffic from the infected approaches and fairway.

Dollar spot symptoms on individual grass blades appear as bleached-out or light tan lesions that may involve the entire width of the blade. Reddish-brown bands are present at the outer edge of the leaf lesions in bentgrass, Kentucky bluegrass, fine-leaf fescue, zoysiagrass, and bermudagrass. The reddish-brown banding does not occur on annual bluegrass.

Occurrence

Dollar spot occurs when the temperature is between 60° and 90°F (15° and 32°C). There appear to be at least two strains; one which occurs during cool weather (when the temperature is below 75°F [24°C]), and one which is favored by high humidity, warm days, and cool nights. Endo et al.[50] and Endo and Malca[49] found a toxin associated with discoloration and deterioration (stunting and necrosis) of bentgrass roots. Endo[48] was able to associate the lack of pathogenicity at 50° and 90°F (10° and 32° Celsius) with lack of toxin production. Toxin production occurred at temperatures between 60° and 80°F (15.5° and 26.8°C), and so did stunting and browning of the roots and the invasion of foliar tissue by the pathogen. Stunting, browning, and necrosis also occurred when the plant roots were separated from the fungus by a cellophane bag that allowed diffusion of the water-soluble toxin only, and the fungus could not subsequently be isolated from the affected roots.

MANAGEMENT

Cultural

Cultural management of dollar spot involves regulating the nitrogen level in the soil. According to one school of thought, the number of infections will be greater at high nitrogen levels but the damage will be less severe than if nitrogen levels are low, because although fewer spots appear in the latter case, they tend to be larger and the damage is more severe. From a practical point of view, during periods of severe dollar spot infection, the nitrogen level must be kept up in order to reduce the severity of the disease and to make the fungicide program more effective. Light and frequent applications of nitrogen are best, both for managing the disease and for general growth of the grass plant, especially during periods of warm weather.

Couch and Bloom[25] found low soil moisture, not nitrogen level, to be important in the development of dollar spot. Their laboratory experiments showed foliar blighting in Kentucky bluegrass grown under conditions of low soil moisture to be more than double that in plants grown at field capacity. Howard[75] and Smith[134] have reported more dollar spot in seasons with low rainfall. Therefore, the turf area should be kept near field capacity and adequate nitrogen provided when dollar spot is a problem. The greatest benefit can be obtained from a good fungicide program when it is combined with these cultural practices.

"Removing the dew" to prevent dollar spot is a common practice on golf course greens. It involves "poling" or "wiping" the dew with a bamboo pole, dragging a hose over the area, or washing the dew off with water. The latter practice may seem to defeat the purpose, since you are trying to dry off the green, but what you really want to do is remove the guttation water which makes up a large portion of what is referred to as dew. Guttation water is a fluid rich in carbohydrates and

Table 2.1. Turfgrass Varieties Most Susceptible to Dollar Spot

Turfgrass Species	Cultivar
Kentucky bluegrass	Nuggett, Sydsport
Red fescue	Dawson
Perennial ryegrass	Manhattan
Bermudagrass	Ormond, Tifway, Sunturf
Zoysiagrass	Emerald
Bahiagrass	Pensacola

amino acids that is exuded from the tip of the grass blade through a structure called a hydrothode when turgor pressure builds up in the plant. This fluid supplies a nutrient-rich medium for the fungus to grow in as it spreads from leaf to leaf. Therefore, anything that will remove guttation water, such as diluting it with irrigation water or breaking up the droplets so the foliage will dry faster, will help reduce the severity of the disease.

Genetic

There are no cultivars of creeping bentgrass that are resistant to dollar spot, and no resistant cultivars of annual bluegrass are known. There are references which list one variety of creeping bentgrass as being moderately resistant or moderately susceptible, but from a practical point of view all are susceptible, and a fungicide treatment will be required in order to prevent serious damage to the turf. For the other grasses, it is simpler to name the most susceptible varieties, so they can be avoided where dollar spot is a major problem (see Table 2.1).

Biological

Nelsen and Craft[106,107] have shown suppression of dollar spot under field conditions with both composted materials like Ringers Compost Plus as well as with a bacterium called *Enterobacter cloacae*. Goodman and Burpee[61] were able to suppress dollar spot in the greenhouse with a fungus called *Fusarium heterosporum*. We have also been able to suppress dollar spot symptoms with Ringers Compost Plus. It gives satisfactory management during years of light to moderate dollar spot infection. However, during years of heavy infections, fungicide treatments will be necessary to obtain satisfactory management.

Chemical

Dollar spot can be managed by many contact and systemic fungicides. For specific chemicals, see Chapter 6 and the Appendix, and for a treatment schedule for specific turfgrasses, see Chapters 11 and 12.

DISEASE: *Pythium* blight (grease spot, cottony blight)

Pathogens: *Pythium aphanidermatum* (Edson) Fitzpatrick (syn. *butleri* subrm.) and *Pythium ultimum* Trow

Hosts: Annual bluegrass, Kentucky bluegrass, bermudagrass, colonial bentgrass, creeping bentgrass, tall fescue, fine-leaf fescues, rough bluegrass, perennial ryegrass, annual ryegrass, redtop

Oogonia of *Pythium*

Pythium blight can be devastating. When the temperature and humidity are high, it can destroy large areas of turf overnight. *Pythium* blight is the only turfgrass disease that you can actually see spreading. No wonder many superintendents consider it the most important disease of golf course turfs—it can kill a lot of grass in a very short time.

Symptoms

The first symptoms of *Pythium* blight are circular reddish-brown spots in the turf, ranging in diameter from 1 to 6 inches (2.5 to 15.3 cm). In the morning dew, infected leaf blades appear water-soaked and dark, and may feel slimy. As they dry, the blades shrivel and turn reddish-brown. When the spots are wet with dew, you can see the active mycelium of the fungi (purplish-gray or white and cottony, depending on the species of *Pythium*) on the outer margins of the spots.

The infected grass plants collapse quickly. If the temperature and humidity remain high, the spots may coalesce, and large areas of turf can be lost in as little as 24 hours. *Pythium* blight follows drainage patterns, and it can be spread by equipment; hence, it often appears in streaks.

Pythium blight is a spot disease. If you notice large, irregular dead patches in your turf without having seen spots first, the problem is not *Pythium* blight, and you must look for a different cause.

Occurrence

Like *Rhizoctonia* and *Colletotrichum, Pythium* is a good saprophyte. It is usually present in the thatch, the soil or both, and requires only the proper conditions to become pathogenic. *Pythium* is also a "water mold" and can survive quite well under water-logged conditions and in the debris of ponds. Ponds are an important source of inoculum, especially in southern areas where pond water is used for irrigation. Each time a golf course is watered with infested pond water, the green, tee, or fairway is reinoculated with *Pythium*, and the disease becomes harder to control.

Although *Pythium* blight is a warm-weather disease that is most destructive when the temperature is between 85° and 95°F (29.4° and 35°C), it is primarily a disease of cool-season grasses. It will attack bermudagrass, especially common bermudagrass, but the damage is less than when it infects cool-season grasses. *Pythium* blight is a prob-

Figure 2.4. ***Pythium*** **blight streaking on a perennial ryegrass fairway.**

lem mainly in the South, where greens, tees, and fairways of dormant bermudagrass are overseeded in the fall with annual ryegrass or perennial ryegrass, which are very susceptible (see Figure 2.4). It is also a serious summertime problem on bentgrass greens in the South. Bentgrasses need constant moisture and, as mentioned above, watering with infested pond water reinfects the grass.

MANAGEMENT

Cultural

Good soil drainage is essential for effective management of *Pythium* blight. Turfgrasses growing in poorly drained or compacted soil have short root systems and, in warm weather, require a great deal of water to prevent wilting. Heavy waterings will increase the severity of *Pythium* blight. Where it is too expensive to rebuild the greens or the tees, or if large parts of the fairways are affected, light watering after sunrise and syringing at midday will help to manage *Pythium* blight. Nighttime waterings should be avoided altogether. There is, however, no substitute for good soil drainage. In its northern range, where it is a problem for just a couple of weeks a year (and not every year), *Pythium* blight occurs only in poorly drained areas. Rebuilding greens in those areas is well worth the time and money. Good air circulation also helps to manage the disease.

High nitrogen levels cause lush growth and make *Pythium* blight worse. No special program of nitrogen fertilization is necessary, since low to moderate nitrogen is normally recommended for turfgrass during hot humid weather.

Genetic

There are no cultivars of creeping bentgrass, colonial bentgrass, annual ryegrass, or perennial ryegrass that are resistant to *Pythium* blight. Of all the turfgrasses, the perennial ryegrasses are the most susceptible to *Pythium* blight, and the fine-leaf fescues and the Kentucky bluegrasses are the least susceptible. Most of the improved bermudagrass cultivars are tolerant of, or resistant to, *Pythium* blight.

Chemical

There are many good fungicides for the control of *Pythium* blight, like the contact fungicides chloroneb, and ethazole, which normally have 5- to 7-day residual, and the systemic fungicides propamocarb, metalaxyl, and fosetyl Al, which normally have a 10–14 day residual when disease pressure is heavy. They can be applied on a preventive basis or using a disease predictor model developed by Nutter et al.[108] and modified by Shane. It requires a maximum daily temperature greater than 82°F (27.7°C) and a minimum temperature of 68°F (20°C), with at least 9 hours of relative humidity above 90%. Predictive models are obviously a more environmentally and economically sound way of applying fungicides, since the fungicide is applied only when the environmental conditions for disease development are present.

What about the old model of adding the temperature and the humidity together to equal 150? This is an excellent model if you are in the business of selling fungicides, because it overpredicts. This means you apply unnecessary fungicide treatments, because this model falsely predicts *Pythium* blight outbreaks when they don't occur. In our studies, this 150 model predicted seven *Pythium* blights one summer when not one single spot of *Pythium* blight occurred. The Nutter-Shane model correctly predicted no *Pythium* outbreaks that season. For those of you who may be too busy in the summer to collect all the data necessary to enter into the model, the Neogen Corporation has developed an instrument called an EnviroCaster that contains the model. It is a mini-weather station that automatically collects all the data, processes it, enters it into the model, and has a little screen that tells you when you need to spray.

Resistance to metalaxyl has been reported by Sanders.[127] Her studies show that the best way to prevent resistance from occurring is to apply 1/2 rate combinations of metalaxyl-propamocarb, metalaxyl-fosetyl Al, or propamocarb-fosetyl Al. Or apply a three-way mixture of 1/3 rates of metalaxyl-propamocarb-fosetyl Al. See Chapter 6 for a complete discussion on delaying fungicide resistance.

DISEASE: Copper spot

Pathogen: *Gloeocercospora sorghi* Bain and Edgerton

Host: Bentgrass (especially low-cut velvet bentgrass)

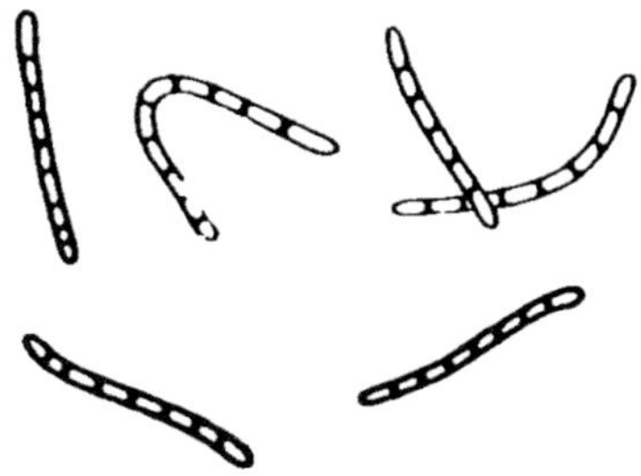

Gloeocercospora sorghi
conidiospores

Copper spot, unlike dollar spot and brown patch, is not a widespread disease on golf course greens, but it can be important when it does occur. It is usually a warm-weather disease (80°F [26.6°C] or higher), although it may begin when temperatures are in the low 70s (21.1 to 23.8°C).[137] It is not unusual to see grass infected simultaneously with dollar spot and copper spot. This disease is often confused with *Microdochium* patch, which occurs at much cooler temperatures.

Symptoms and Occurrence

Gloeocercospora sorghi survives periods away from the host as sclerotia in the mat and thatch. Infection can take place rapidly (within 24 hours) and considerable turf may be lost with heavy infections. The leaf lesions are small and reddish at first. They enlarge to cover the entire width of the blade, and can eventually destroy the entire leaf blade. The spots that develop in the turf are usually 1 to 3 inches (2.5 to 7.6 cm) in diameter and are copper to reddish pink. The pink color is due to the pink spore masses of the fungus, which are gelatinous when observed in the early morning "dew."

MANAGEMENT

Cultural

Avoid applying excess nitrogen prior to and during the time copper spot is a problem. The disease is worst on low pH soils. Liming may help reduce the severity of the disease.

Genetic

No information is available on cultivars resistant to copper spot.

Chemical

Chlorothalonil and anilazine will manage copper spot if it occurs. Since fungicides are normally part of a good preventive program, a special spray program for copper spot should not be necessary. However, the timing of fungicide application may have to be altered where copper spot is a problem.

DISEASE: Yellow tuft (downy mildew)

Pathogen: *Sclerophthora macrospora* (Sacc.) Thirum, Shaw, and Naras

Hosts: Kentucky bluegrass, creeping bentgrass, annual bluegrass, fine-leaf fescues, tall fescue, perennial ryegrass, *Poa trivialis*, colonial bentgrass, velvet bentgrass

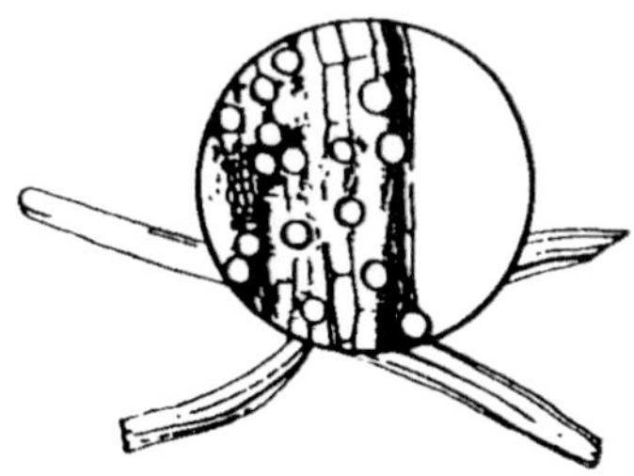

Yellow tuft

In the 1950s, Tarjan and Hart associated yellow tuft with parasitic nematodes[144] but this was never verified. Mueller et al.[105] found the downy mildew fungus *Sclerophthora macrospora* associated with yellow tuft, and Jackson and Dernoeden[33,79] later demonstrated that *S. macrospora* is the causal agent.

Symptoms

The symptoms are small yellow spots in the turf, from 1/4 to 3 (.6 to 7.6 cm) inches in diameter. Infected turf may appear spotted, as though droplets of herbicide had been misapplied to the area. On a new creeping bentgrass green it looks like an invasion of annual bluegrass. Examination of the tufts reveals a dense cluster of yellow shoots, which can be seen arising out of a single axillary bud at the node or terminals of a stem.[79]

Occurrence

Yellow tuft most commonly occurs in the cool weather of the spring and fall. *S. macrospora* can overwinter and oversummer as mycelium in the infected plant or as zoospores in grass plants and in plant debris. With the advent of cool moist weather, mycelia in infected plants produce sporangia which germinate and release motile zoospores that can infect healthy plants. Infected plants have poorly developed root systems, and die during stress periods. It is not a devastating disease but one that, over time, will eliminate the creeping bentgrass you planted and allow annual bluegrass to fill in the voids. The damage is greatest in areas where surface drainage occurs on a green.

MANAGEMENT

Cultural

The addition of iron (Fe_2SO_4) will help to mask the symptoms. No more than one-half pound (.25 kg) of nitrogen per 1000 square feet (100 square meters) per month should be applied in June, July, and August, and it may be even better to make no nitrogen applications at all.

Chemical

The application of metalaxyl prior to symptom development will manage the disease. Once the symptoms have developed, the hormonal response, (formation of the tuft) cannot be reversed. This means applications need to be begin following the first mowing and continue until the nighttime temperatures remain above 65°F (18.3°C) on a consistent basis. Fungicide applications should begin again in the fall, when the nighttime temperatures drop into the 50s (under 10°C).

CHAPTER 3

Fungal Diseases of Turfgrass II: Diseases Primarily of Home Lawns and General Turfs

THE *HELMINTHOSPORIUM* DISEASES

The *Helminthosporium* diseases or "leaf spot," as they are commonly referred to, are very widespread on both cool- and warm-season turfgrass species. Even though the genus is no longer officially referred to as *Helminthosporium*, the term still persists among lay people. Therefore, they will still be referred to as the *Helminthosporium* diseases when discussing them in general, with the correct scientific name being used for the specific diseases. The common names of the *Helminthosporium* diseases, the scientific names of the organisms, and their hosts are listed in Table 3.1.

Table 3.1 The *Helminthosporium* Diseases and Their Hosts

Common Name Affected	Pathogen	Turfgrasses
Leaf spot	*Bipolaris sorokiniana* (syn. H. sativum)	Annual bluegrass Kentucky bluegrass, colonial bentgrass, creeping bentgrass, perennial ryegrass, fine-leaf fescue, meadow fescue, tall fescue
Melting-out	*Drechslera poae* (syn. H. vagans)	Kentucky bluegrass
Net-blotch	*Drechslera dictyoides* (syn. H. dictyoides)	fine-leaf fescues, tall fescue
Red leaf spot	*Drechslera erythrospila* (syn. H. erythrospilum)	creeping bentgrass
Brown blight	*Drechslera siccans* (syn. H. siccans)	perennial ryegrass annual ryegrass
Leaf blotch	*Bipolaris cynodontis* (syn, H. cynodontis)	bermudagrass
Zonate eyespot	*Drechslera gigantea* (syn. H. giganteum)	bermudagrass

DISEASE: Melting-out

Pathogen: *Drechslera poae* Shoemaker (syn. *Helminthosporium vagans* Drechslera).

Host: Kentucky bluegrass

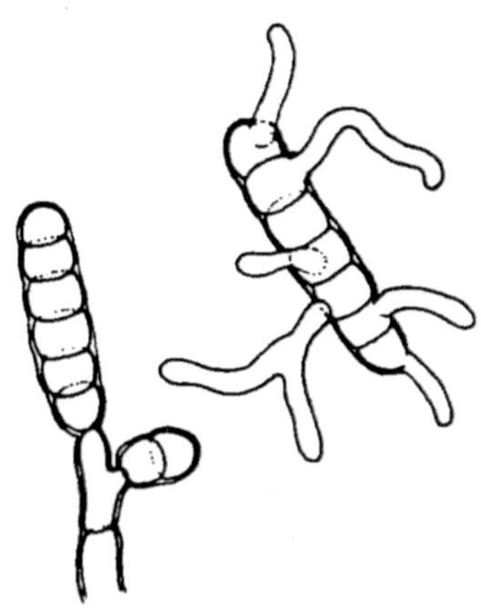

Drechslera poae
conidiophore and conidiospores
(one germinating)

Melting-out is potentially the most important disease of Kentucky bluegrass. This may surprise many people who think of summer patch or necrotic ring spot as the most serious disease affecting Kentucky bluegrass turf. The reason patch diseases are so important today is because there are now so many Kentucky bluegrass cultivars developed for resistance to melting-out. Merion was the first; hence, its popularity. Before Merion was developed, the main drawback of Kentucky bluegrass was its susceptibility to melting-out. In other words, if there were no cultivars resistant to melting-out, summer patch or necrotic ring spot would not be a problem because melting-out would destroy the turf long before they appeared.

This brings up an important aspect of selecting or breeding for resistance to a particular pathogen or disease. The introduction of varieties resistant to a major disease often results in the emergence of diseases previously minor or unknown in that turf species. For example, before the selection of Merion, summer patch or necrotic ring spot was unknown, and stripe smut and rust were unimportant in Kentucky bluegrass. Red thread was unknown as a disease of perennial ryegrass before the introduction of Manhattan.

Helminthosporium diseases of Kentucky bluegrass are usually referred to as either "melting-out" or "leaf-spot." The pathogens responsible for these diseases are *Drechslera poae* and *Bipolaris sorokiniana*, respectively. *D. poae* is a cool-weather pathogen, whereas *B. sorokiniana* is a warm-weather pathogen.

Symptoms

The first symptoms of melting-out are black to purple spots on the leaf blades. Spots eventually appear on the leaf sheaths, and the fungus moves down and invades the crowns and roots of the grass plant. From a distance, the turf will appear either yellow or blackish-brown. These colors are a reflection of the nitrogen level of the turf during infection. At low nitrogen levels, infected turf turns yellow; with high nitrogen levels, it turns blackish-brown. As the disease progresses, the infected area becomes thin and brown, similar in appearance to wilted turf. Unless a fungicide is applied, what was once a Kentucky bluegrass turf will die and be replaced by broadleaf weeds and weedy grasses.

Occurrence

Melting-out is described by Britton[13] as a disease that has a leaf-spot stage in the cool, wet period of spring or fall, followed by a melting-out or a crown-and-root-rot stage in the warm weather of summer. In my opinion, it is a cool-weather disease, and Monteith and Dahl[104] agree. Why the discrepancies? I believe we are dealing with two separate diseases: (1) melting-out, caused by *D. poae*, which is a problem during cool weather, and (2) leaf spot, caused by *B. sorokiniana*, which has a melting-out stage that occurs during warm weather.

The leaf-spot stage of melting-out begins early in the spring with the coming of cool wet weather (especially nice, slow drizzle). The melting-out or crown-and-root-rot stage of the disease will follow, and the infected Kentucky bluegrass will begin to thin and die. Severe loss of turf can occur within a 3-week period. Then, with the coming of warm dry weather (temperatures above 70°F [21.1°C]), the infected turf will begin to recover and fill in. The degree of recovery depends on several factors, including the severity of the infection, the varieties of Kentucky bluegrass, the nitrogen level, and the irrigation practices. The life cycle of *D. poae* is illustrated in Figure 3.1.

Fall infection with *D. poae* should also be mentioned. Such infections do occur, but the melting-out stage is rarely seen unless protracted cool, wet weather occurs. This does not mean that fall infection should be ignored or considered unimportant, since it can give rise to turf being lost under snow or early loss of turf in the spring, sometimes before any fungicide applications can be made.

MANAGEMENT

Cultural

Mowing Height

Raising the height of cut during the cool, wet weather of spring and fall will help the turfgrass stand to survive an attack of melting-out. Although this may not be practical on a golf course fairway, it will certainly benefit a home lawn. Too often Kentucky bluegrass suffers from what I call golf course fairway syndrome. It occurs when someone admires the appearance of golf course fairways they have been playing on and decides they want their home lawn to look that way. So they proceed to lower the lawn mower to its lowest setting to bring the lawn down to the mowing height of the fairway. The problem is that the grasses on the fairways are creeping bentgrass and annual bluegrass which can tolerate the low height of cut. The Kentucky bluegrass will not. Kentucky bluegrass grows best at between 2 to 3 inches (5 to 7.6 cm), and does not survive well below that height of cut. If you don't

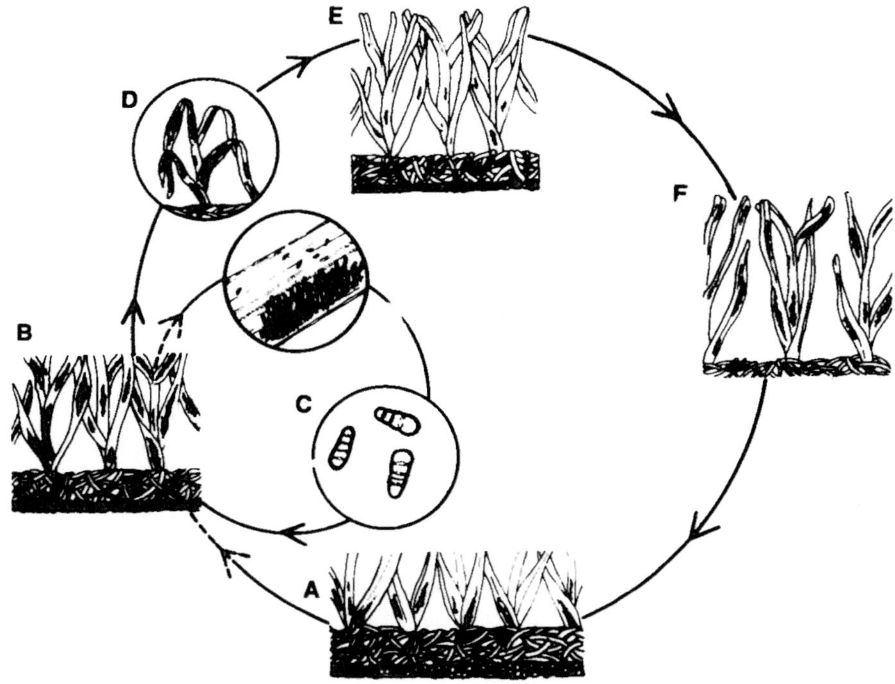

Figure 3.1. **Melting-out of Kentucky bluegrass caused by *Drechslera poae*. (A) *Drechslera poae* overwintering on the lower portions of the plants and on dead plant material in the thatch; (B) severe infection of the foliage during the cool wet weather of early spring; (C) sporulation and reinfection continue as long as the cool wet weather persists. The crowns and root are affected also; (D) severe thinning of the turf and death of the grass plant result in susceptible cultivars if the cool weather persists for more than 2 weeks; (E) grass plants that are still alive recover with the arrival of warm weather. *D. poae* oversummers mainly in the surface layer of the thatch and on the lower parts of the plants; (F) foliar infection begins again with the advent of cool, wet, fall weather. Severe thinning usually does not occur at this time, but infections can lead to severe loss of turf under snow cover in the spring, or to severe epidemics when the turf greens up.**

believe that, contact your local superintendent who watched his 3/4 inch (2 cm) Kentucky bluegrass fairways turn into annual bluegrass fairways.

Fertility

Applying adequate nitrogen levels during the cool weather of spring and fall will help to reduce the severity of the disease. Although this is contrary to conventional beliefs, field research has shown dormant nitrogen applications and moderate spring application will reduce the severity of the disease under field conditions.[155] The original work showing the increased susceptibility of Kentucky bluegrass to melting-out was in the greenhouse. So, everyone managing Kentucky bluegrass in the greenhouse: don't apply any nitrogen to the turf. For the rest of you, who are maintaining Kentucky bluegrass out in the field, moderate levels of nitrogen will greatly reduce the severity of the disease!

Irrigation

Daily irrigation will reduce the severity of melting-out.[155] Colbaugh and Endo found similar results with leaf spot.[22] The turf should receive between 0.10 to 0.20 of an inch (.25 to .5 cm) of irrigation every day. The ideal time to apply this irrigation is midday. In the more northern high humidity climates, .10 of an inch (.25 cm) should be adequate, and the 0.20 of an inch (0.5 cm) is for the more arid areas. The combination of nitrogen fertility and irrigation should allow the disease to be managed culturally, avoiding the need for fungicides and fitting well into IPM programs.

Genetic

See Chapter 8 for a discussion of the subject.

Chemical

See Chapter 6, the Appendix, and page 43.

DISEASE: Leaf spot

Pathogen: *Bipolaris sorokiniana* Sacc. ex Sorokin (syn. *H. sativum* P.K.B.; *Drechslera sativus* Drechslera ex. Dastur)

Hosts: Annual bluegrass, Kentucky bluegrass, fine-leaf fescues, tall fescue, colonial bentgrass, creeping bentgrass, perennial ryegrass

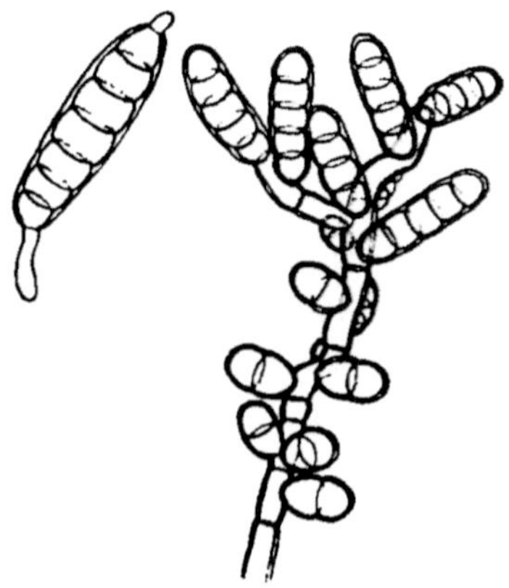

Bipolaris sorokiniana

Bipolaris sorokiniana is the most widely occurring of the *Helminthosporiums*. It attacks Kentucky bluegrass, fine-leaf fescues, tall fescue, colonial bentgrass, creeping bentgrass, annual bluegrass, and perennial ryegrass.

Symptoms

The first symptoms are small dark purple to black colored spots on the leaf blade (Figure 3.2). As the spots enlarge, the centers often turn light tan. At temperatures exceeding 85°F (30°C), distinct spots are often absent and the entire blade appears dry and straw-colored. The disease usually is confined to the leaf blades during cool weather. Leaf sheaths, crowns, and roots will become infected during hot, humid

Figure 3.2. Close-up of a melting-out lesion on Kentucky bluegrass leaf.

weather, and severe thinning of a turfgrass stand can occur in a short period of time.

Occurrence

Leaf spot is a warm-weather disease, as is red leaf spot (net-blotch, brown blight, and melting-out are cool weather diseases). *B. sorokiniana* overwinters as dormant mycelium in infected plants and in the dead grass debris that makes up the mat. When the temperature reaches 70°F (21.1°C) or more, the leaf spot phase of the disease becomes evident. At 85°F (29.4°C) the leaf-blighting stage begins and becomes progressively worse as the temperature rises. The disease is most severe when the temperature reaches 95°F (35°C) and the humidity is high.[167] Colbaugh and Endo[22] showed the disease to be most severe under conditions of drought stress followed by rewetting. They believed that the germination of *B. sorokiniana* spores increased during the period of drought, owing to the absence of a volatile inhibitor which is present when the mat remains moist. The inhibitor returns an hour after rewetting.

MANAGEMENT

Cultural

Moderate levels of nitrogen should be applied during warm weather, when the leaf spot is active. Potassium and phosphorus should be maintained at adequate levels as determined by soil tests. The work of Colbaugh and Endo[22] suggests that keeping the turf moist through daily irrigation during periods when *B. sorokiniana* is active should help to reduce or even prevent development of the disease.

Genetic

There are very little data on the resistance to leaf spot of Kentucky bluegrass, colonial bentgrass, creeping bentgrass, fine-leaf fescues, tall fescue, Italian ryegrass, and perennial ryegrass. That is not to say that there are no resistant cultivars. There may well be some, especially among the new improved Kentucky bluegrasses and perennial ryegrasses, but none have been idenfified yet.

Chemical

Both contact and systemic fungicides can be used to manage the *Helminthosporium* diseases. The contact fungicides, anilazine, chlorothalonil and maneb, must be applied every 7 to 10 days to be effective, whereas the systemic fungicides, iprodione and vinclozolin, can be applied on a 14- to 21-day schedule. For the best results, apply the fungicides preventively just before the disease season begins. For example, to prevent melting-out in Kentucky bluegrass, apply the fungicide when the temperatures are between 50° and 60°F (10° and 15.5°C), during a rainy period. For leaf spot on the fine-leaf fescues, apply the fungicide just before the weather turns warm. For more information on the fungicides available to treat *Helminthosporium* diseases, see Chapter 6 and the Appendix; for appropriate times to apply, see Chapters 11 and 12.

DISEASE: Net-blotch

Pathogen: *Drechslera dictyoides* Paul and Parbery.(syn. *Helminthosporium dictyoides* Drechslera)

Hosts: Fine-leaf fescues, tall fescue, meadow fescue

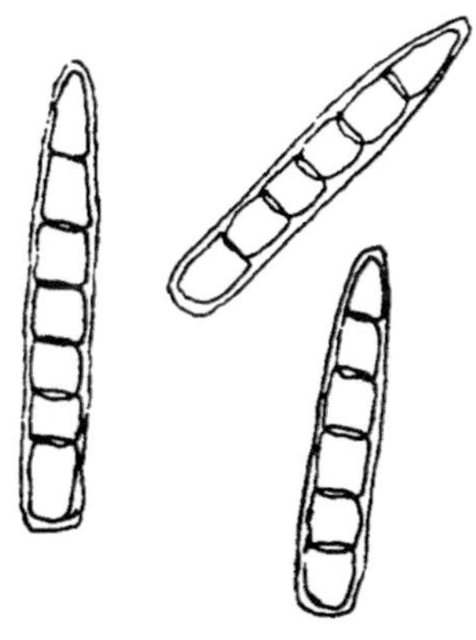

Drechslera dictyoides conidiospores

Net-blotch is a disease that occurs during the cool weather of spring and fall. It is often called leaf spot, but leaf spot is caused by *Drechslera sorokiniana*. Since the primary problem on fine-leaf fescues occurs in the warm weather of the summer, perhaps a lot of what was thought to be 'leaf spot' caused by *Drechslera dictyoides* is actually leaf spot caused by *B. sorokiniana*. Numerous isolations from *Helminthosporium* lesions have been made for Michigan State University's fine-leaf fescue breeding program, and over 90% of them have been *B. sorokiniana*, regardless of the time of year.[159] Data from the southern half of Michigan obviously do not imply that this is the case everywhere, but our findings do suggest that leaf spot, not net-blotch, may be the important disease on the fine-leaf fescues.

Symptoms

The initial symptoms are similar to the other *Helminthosporium* diseases—tiny purplish to black colored spots on the leaf blade that eventually enlarge. On the fine-leaf fescues the enlarged spots usually extend the entire width of the blade. On the broad-leaf species, like the tall fescues and meadow fescue, the disease initially occurs as irregular dark purple to black transverse strands, resembling dark threads drawn across the leaf. These strands look like a net; hence, the name net-blotch.

Occurrence

The development of net-blotch is described in the literature as similar to *Helminthosporium* melting-out in Kentucky bluegrass. The fungus survives the dormant period as mycelium in infected tissue on the living host and in the mat. In the spring, spores are produced in infected plant parts, and are water splashed or wind blown to healthy plant tissue. The leaf-spot phase of the disease is supposedly inconspicuous in the spring. Crown and root infection are thought to take place during this time, although the symptoms are not expressed until the warm dry weather of summer. In actual fact, both the leaf-spot and the crown- and root-rot phases take place during the cool weather of spring and fall. The decline of fine-leaf fescue during the warm weather is due to *B. sorokiniana*, which is readily isolated from the fine-leaf fescues during the warm weather of summer. It appears in the cooler regions of the world where oceanic climates exist, i.e., the Pacific Northwest, Atlantic Northeast, the British Isles, northern Europe; the major pathogen on the fescues is *D. dictyoides*. The major pathogen on the fescues in the more temperate climates of the world is *B. sorokiniana*. Both pathogens limit the widespread use of fine-leaf fescues. Finding cultivars resistant to these pathogens should allow much wider use of the fine-leaf fescues and extend their range.

MANAGEMENT

Cultural

The fine-leaf fescues are grasses that grow best with low maintenance. Using minimal amounts of nitrogen, with normal levels of phosphorus and potassium, will help fine-leaf fescue stands survive. Minimal amounts of water should also be used. The fine-leaf fescues can withstand a tremendous amount of drought stress, but they cannot stand overwatering. The tall and meadow fescues also require benign neglect for survival.

Genetic

There are no cultivars of fine-leaf fescue or any other fescue species that are truly resistant to either *B. sorokiniana* or *D. dictyoides*. In general, the hard fescues have shown more resistance to both pathogens than either the creeping red fescues or the chewings fescues.

Chemical

See Chapter 6, the Appendix, and page 43.

DISEASE: *Helminthosporium* brown blight

Pathogen: *Drechslera siccans* Shoemaker (syn. *Helminthosporium siccans* Dreschler)

Hosts: Italian ryegrass, perennial ryegrass

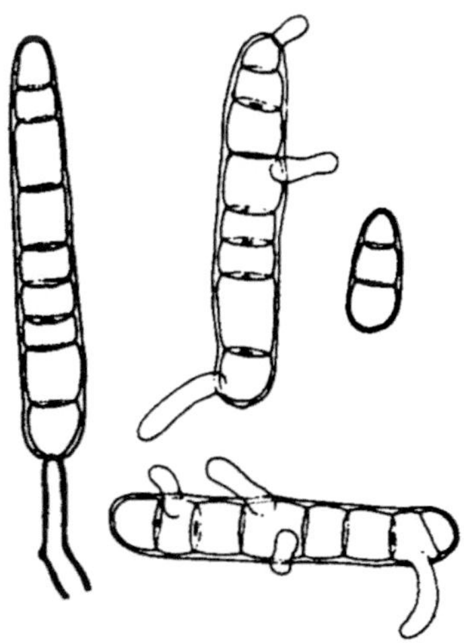

Drechslera siccans
conidiospores (two germinating)

Brown blight is a serious problem on ryegrass during the cool weather of spring and fall. It is becoming one of the factors limiting the successful adaptation of some of the new perennial ryegrasses, most of which were selected for their finer texture, better mowability, and other agronomic characteristics, with little regard for potential disease problems.

Symptoms

The first symptoms are small brown spots that appear on the grass blades. The spots enlarge and may develop white centers. A second foliar symptom consists of dark brown streaks on the leaf blade. Both symptoms may occur simultaneously on the same blade. The entire blade of susceptible varieties can be destroyed if the infection is severe. Turf that is heavily infected with brown blight may show extensive thinning.

Occurrence

Like the other *Helminthosporiums*, *Drechslera siccans* survives the dormant period in the infected tissue of living plants and in the dead tissue making up the mat. In the spring, spores are produced in the infected tissue and are wind blown and water splashed to healthy tissue. Brown blight is supposed to have a leaf-spot stage in the spring, followed by rotting of the root and crown in the summer and another leaf-spot stage in the fall. However, much as in melting-out in Kentucky bluegrass, what actually happens is that the whole process (leaf-spot stage, crown-and-root-rot stage) occurs during the cool weather of the spring and fall, with few signs of disease evident during the warm summer months.

MANAGEMENT

Cultural

Adequate levels of nitrogen during cool weather will reduce the severity of brown blight. Red thread and crown rust are also a problem during cool weather, and the addition of nitrogen will also help manage these two diseases. Moreover, the time of year when brown blight is a problem is also the time of year when the plant can best utilize nitrogen for its growth and development.

Genetic

There are varieties of perennial ryegrass that are resistant to brown blight, and they should be used in preference to the susceptible varieties. Check with your local turf expert for the resistant cultivars best adapted for your area. Some of the cultivars that are resistant to brown blight may be susceptible to other serious diseases like brown patch and red thread. The situation is similar to that in Kentucky bluegrasses: many varieties are resistant to melting-out, but they shouldn't be used because of their susceptibility to necrotic ring spot, summer patch, or stripe smut. You need to determine which of the various diseases is the most serious or most difficult to manage, and choose cultivars that are the most resistant to that disease and manage the other diseases chemically. (For a more detailed discussion, see Chapter 11.) The resistance and susceptibility of the perennial ryegrasses to various diseases are discussed in Chapter 9.

Chemical

See Chapter 6, the Appendix, and page 43.

DISEASE: Leaf blotch

Pathogen: *Drechslera cynodontis* Nelson. (syn. *Helminthosporium cynodontis* Marig.) *H. giganteum* Heald and Wolf, *H. rostratum* Drechsl., *H. spiciferum* (Bain.) Nicot, *H. stenospilum* Drechsl., and *H. triseptatum* Drechsl. are all known pathogens of bermudagrass

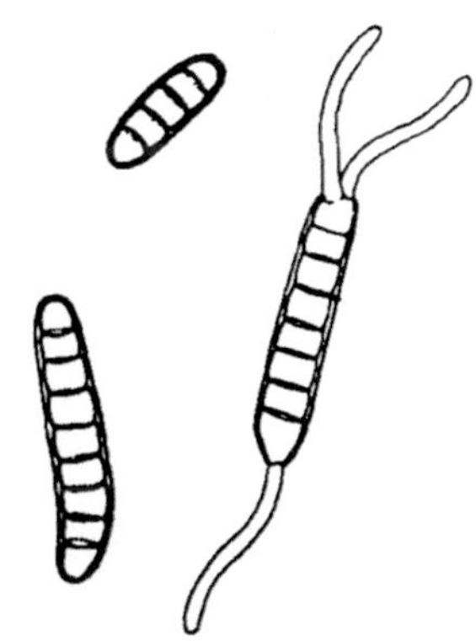

Drechslera cynodontis conidiospores (one germinating)

Host: Bermudagrass

Leaf blotch is caused by *Drechslera cynodontis*. The other pathogens listed may cause diseases that are just as damaging, but there is not sufficient information to warrant separate discussions of them. Therefore, leaf blotch will be discussed as an example of a *Helminthosporium* disease on bermudagrass. Keep in mind, however, that other *Helminthosporium* diseases of bermudagrass may be equally important. Leaf blotch can be a serious problem, especially in humid areas and during years of frequent spring and fall rains.

Symptoms

The first symptoms are small olive-green spots that appear on the leaves. The spots later grow into big brownish-green to black blotches (Figure 3.3). If the infection is severe, the leaves eventually wither and

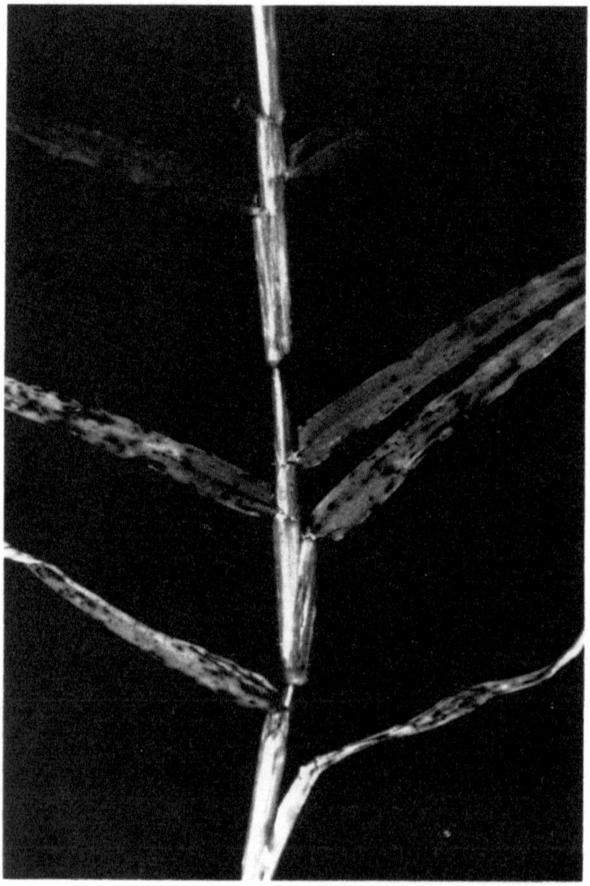

Figure 3.3. Leaf blotch on bermudagrass. (Photograph courtesy of Dr. Don Blasingame.)

become light tan. Turf infected by *D. cynodontis* is straw-colored. The affected area may be small at first and then spread.

Occurrence

The fungus survives as dormant mycelium in the infected tissue on the host and in the mat. With the advent of wet weather in the late winter and early spring and the breaking of dormancy in the bermudagrass, the fungus begins to sporulate. The spores are then wind blown and water splashed to healthy tissue, which they infect. Leaf blotch tends to disappear with warm weather, much like melting-out of Kentucky bluegrass, but it may return with the cool, wet weather in late summer and early fall. Thus, bermudagrass may enter the dormant period in a weakened condition and die during the winter.

MANAGEMENT

Cultural

Adequate nitrogen in the early spring and early fall will help to manage leaf blotch. A balanced program of phosphorus and potassium should be maintained. Avoid excess watering during the early spring and fall, or water in the daytime so the foliage has a chance to dry.

Genetic

No information is available on the resistance of bermudagrass cultivars to leaf blotch.

Chemical

See Chapter 6, the Appendix, and page 43.

DISEASE: Red leaf spot

Pathogen: *Drechslera erythrospila* Shoemaker (syn. *Helminthosporium erythrospilum* Drechsler)

Hosts: Redtop, colonial bentgrass, creeping bentgrass, velvet bentgrass

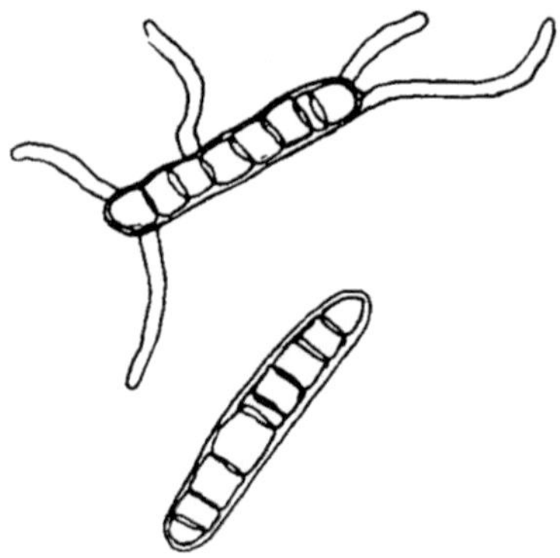

Drechslera erythrospila conidiospores (one germinating)

Symptoms

Red leaf spot is a disease of the bentgrasses and redtop that usually occurs during warm wet weather. The lesions start out similar to other *Helminthosporium* lesions as small, dark reddish-brown spots on the leaf blade. As the spots grow, they develop straw-colored centers surrounded by reddish-brown borders. If warm wet weather continues, the entire blade may wither, giving the stand of turf a drought-stricken appearance. Severe thinning of creeping bentgrass may be caused by red leaf spot infections.

Occurrence

The fungus survives the dormant period in infected tissue of living plants and in the infected dead tissue in the mat. In the late spring, spores are produced in these infected areas and are wind blown or water splashed to healthy bentgrass plants. During periods of warm wet weather, red leaf spot can cause severe thinning of bentgrass turfs. On creeping bentgrass greens, even moderate thinning can be a serious problem. Thirty-percent thinning of a home lawn or a fairway is sometimes tolerable, but on a putting green that amount of thinning is disastrous.

MANAGEMENT

Cultural

Meyer and Turgeon[102] report that high levels of nitrogen, in conjunction with fungicide treatment, provide the best control. It would appear, based on recent findings concerning the effect of nitrogen on the *Helminthosporium* diseases, that moderate levels of nitrogen will help reduce the severity of the disease.

Genetic

There are no cultivars known to be resistant to red leaf spot.

Chemical

See Chapter 6 and the Appendix.

DISEASE: Fading-out, *Helminthosporium-Curvularia* complex

Pathogens: *Curvularia lunata* (Wakker) Boedijn, *C. geniculata* (Tracy and Earle) Boedijn, *C. intermedia* Boedijn, *C. protuberata* Nelson and Hodges

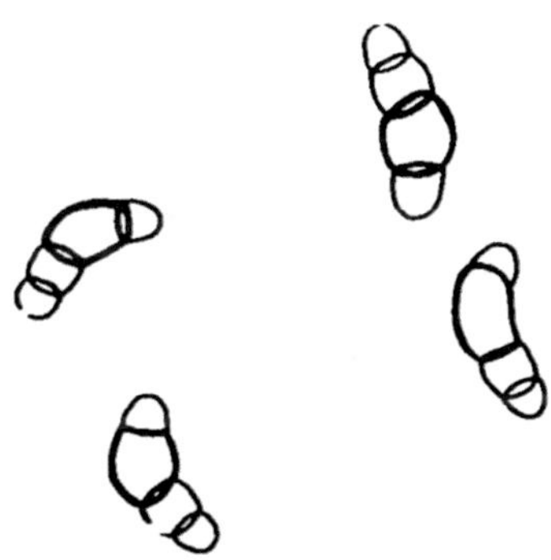

Curvularia conidiospores

Hosts: Annual bluegrass, Kentucky bluegrass, creeping bentgrass, fine-leaf fescue, *Poa trivialis* L., Canada bluegrass

Plant pathologists are still debating whether *Curvularia* causes a disease of turf or whether it is simply a secondary organism (saprophyte) invading an old *Helminthosporium* lesion. Madison[99] tended to discount *Curvularia* as a pathogen, while Howard[75] has supported the idea of *Curvularia* as a pathogen, either alone or as part of a *Helminthosporium-Curvularia* complex. Brown et al.[14] showed that four species of *Curvularia* are pathogenic on Kentucky bluegrass, spreading fescue, and creeping bentgrass. Hodges and Madsen[72] concluded that *Curvularia geniculata* was a weak pathogen on Kentucky bluegrass under high temperature conditions (25° to 35°C), but that it is part of a true disease complex with *Bipolaris sorokiniana* at 30°C. Lesions are initiated by *B. sorokiniana* and then necrotic tissue is invaded by *C. geniculata*, which is extremely aggressive at high temperatures. The occurrence of both pathogens results in a more severe disease than either is capable of causing alone. While the debate will probably go on forever, the evidence put forth by Brown, et al.[14] and Hodges and Madsen[72] leaves little doubt that under proper conditions *Curvularia* species, either alone or in conjunction with *H. sorokinianum*, can be important turfgrass pathogens.

Symptoms

On Kentucky bluegrass and fine-leaf fescue, the grass blades turn yellow from the tip back, eventually becoming brown and then gray, until finally the whole blade looks shriveled. On creeping bentgrass the blades go from yellow to tan to shriveled. Leaf spots with tan centers and red or brown margins are sometimes observed on red fescue and Kentucky bluegrass, but never on creeping bentgrass.[14]

Occurrence

Fading-out is a warm-temperature disease (above 85°F [30°C]). Like anthracnose, it appears to attack grass plants that are experiencing heat stress or high-temperature growth stoppage. The disease is favored also by high humidity. Since *Curvularia* is a good saprophyte, it can survive quite well in the plant debris until the ideal conditions are present, and then it becomes pathogenic.

MANAGEMENT

Cultural

Syringing to cool the microclimate temperature will help reduce the severity of fading-out in open areas, but in areas with poor air circulation it may worsen the disease by raising the humidity.[14]

Genetic

No information is available on cultivars resistant to fading-out.

Chemical

The only chemical labeled for managing fading-out is chlorothalonil.

DISEASE: Powdery mildew

Pathogen: *Erysiphe graminis* DC.

Hosts: Bermudagrass, Kentucky bluegrass, fine-leaf fescues, redtop

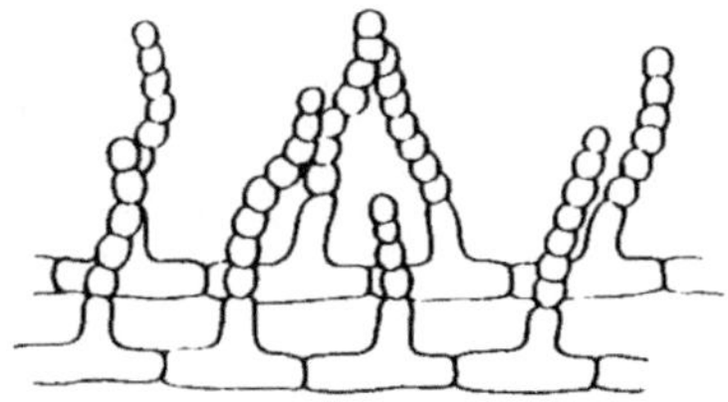

Erysiphe graminis conidiospores

Powdery mildew is a classic example of a disease that attacks turfgrass plants growing under suboptimal environmental conditions. When bermudagrass, Kentucky bluegrass, redtop, and fine-leaf fescues are grown under the sunlight conditions that are optimal for them, they are resistant to powdery mildew. They become susceptible only when they are grown under low light intensity. For example, most tests say that Merion Kentucky bluegrass is susceptible to powdery mildew and resistant to melting-out. Yet when Merion is grown in full sunlight it is resistant to both diseases, and when it is grown in the shade it loses its resistance to them. What ought to be said is that Merion is resistant to both powdery mildew and melting-out when grown in full sunlight, and susceptible to both when grown in the shade.

Symptoms

Turf infected with mildew has a grayish white cast and looks as though it has been dusted with flour. The first symptoms are individual white patches on the leaf blade, but eventually the whole blade becomes covered with one white mass. As the disease progresses, the leaves turn yellow, wither, and die (see Figure 3.4). When the leaves begin to deteriorate, brown to black spore-bearing bodies (cleistothecia) may be found.

Occurrence

Erysiphe graminis overwinters as ascospores in cleistothecia. The ascospores are released in the spring, germinate, and infect turf in the late spring or early summer. In about 4 days the primary infection produces numerous spores (secondary inoculum) which reinfect the

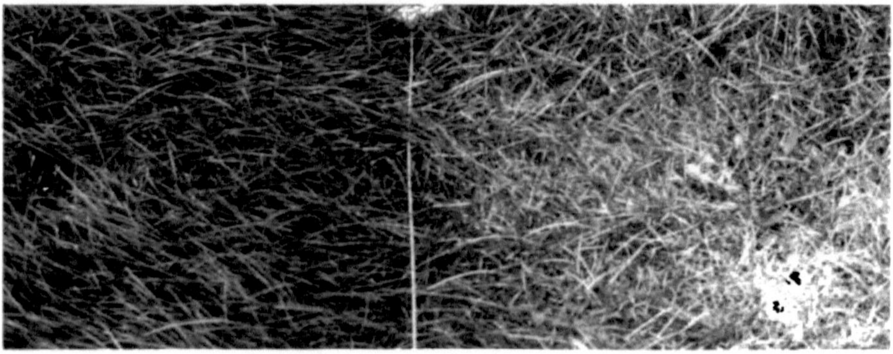

Figure 3.4. Powdery mildew thinning out Kentucky bluegrass stand where no chemical treatment has been applied, compared to the EL-273 treated plot.

grass, and the cycle continues throughout the season as long as there is some turf left.

The conidiospores are clear with thin cell walls, and can germinate in about 2 hours. Although most fungi that infect turfgrass require free water for their spores to germinate, this is not true of *E. graminis*, which can germinate if the humidity is high enough.

An interesting thing about the infection caused by *E. graminis* is that only the epidermal layer of leaf cells is invaded by the fungus. By means of a specialized feeding structure (haustorium), the powdery mildew fungus can extract the nutrition it needs from the living cell. Most fungi that are obligate parasites have haustoria.

MANAGEMENT

Cultural

The obvious way to manage powdery mildew culturally is to remove the shade. But you are not going to cut down your favorite tree just to grow a little grass, and no matter how hard you try, you cannot eliminate the north side of your house. In the case of the tree, however, pruning the low branches will increase air circulation and let in a little more light.

Genetic

The fine-leaf fescues are the grasses usually recommended for shady places. The fine-leaf fescues do fine in open shade but not in dense shade, because they are susceptible to leaf spot and powdery mildew. *Poa trivialis*, perennial ryegrass and annual bluegrass are the only other species that grow well in the shade. Both are considered weeds, but you can mow them, and they both look better than bare ground. In northern areas of the cool-season grass belt, both Nugget and Bensun Kentucky bluegrass have been grown successfully in shaded areas.

Chemical

Chemical management of powdery mildew can be accomplished with the DMI fungicides. The problem is *E. graminis* will eventually develop resistance to this group of fungicides as it did to the benzimidazoles.[153] If you don't like *P. trivialis*, perennial ryegrass or *P. annua*, try ornamental plants.

DISEASE: Pink patch

Pathogen: *Limonomyces roseipellis*

Host: Perennial ryegrass, colonial bentgrass, creeping bentgrass, annual bluegrass, fine-leaf fescues

Pink patch and red thread were thought to be the same disease for a long time. They have only recently been separated into two distinct diseases caused by different pathogens. It is not unusual to find them occurring together.

Symptoms

Pink patch occurs as circular patches in infected turfs. In very susceptible turfs like colonial bentgrass it can produce "frog eye" symptoms similar to those produced by the patch disease fungi. The turf in the infected areas is often dark reddish-brown in color. In less susceptible species, like creeping bentgrass, the patches appear as slightly off-colored turf with a little thinning. Pinkish-colored mycelium can be observed on the infected plants during periods of high humidity or high moisture. The reddish-colored threads and cottony flocks of arthroconidia that are present with red thread are absent. If in doubt as to whether the disease you have is red thread or pink patch, place infected foliage into a jar with a moist paper towel, seal it, and let it remain for 24 hours. If it is pink patch, the mycelium of the fungus appear as miniature cottony puffs on the foliage.

Occurrence

The disease occurs primarily in the cool wet weather of the late summer and fall. It is a low nitrogen disease and consequently is most severe on under-fertilized, slow-growing turfs. It can cause severe damage to colonial bentgrass under these conditions.

MANAGEMENT

Cultural

Having adequate levels of nitrogen applied to the turf throughout the late summer and fall is important in managing pink patch. If turfgrass growth slows down due to the cool weather of the fall rather than lack of nitrogen, fungicide applications may have to be made to manage pink patch, especially on colonial bentgrass. Care should be taken not to encourage *Microdochium* patch with nitrogen in the fall. (See discussion under red thread.)

Genetic

Nothing is known about host resistance to this disease.

Chemical

Although little research is available on fungicides for the management of pink patch, products like chlorothalonil, vinclozolin, iprodione, triadimefon, fenarimol, and propiconazole should manage the disease.

DISEASE: Red thread

Pathogen: *Laetisaria fuciformis* (McAlp.) Burdsall

Hosts: Annual bluegrass, creeping bentgrass, Kentucky bluegrass, perennial ryegrass, fine-leaf fescue, bermudagrass

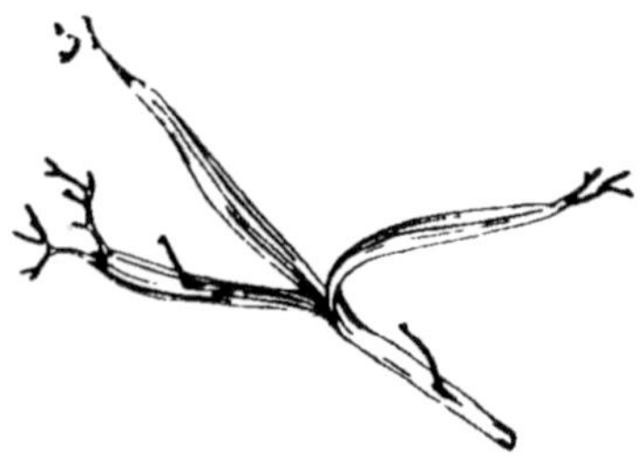

Laetisaria fuciformis stroma on infected leaves

Red thread is a disease of slow-growing turf, whether the slow growth is caused by lack of nitrogen or by cold temperatures. It is a problem in the northern part of the United States and in Canada, New Zealand, Great Britain, and the north coastal areas of Europe, where the cool weather that often slows turf growth is accompanied by frequent drizzle or fog.[137] It has also been reported on bermudagrass.[55]

Red thread has become a greater problem since the trend toward lower nitrogen rates to encourage creeping bentgrass in golf course turfs. Even before the cost of fertilizer went up, it was fairly common to see red thread thinning roughs, which often were maintained with low levels of nitrogen. Now with the lower rates of nitrogen used to encourage creeping bentgrasses, it is not uncommon to find red thread on many fairways as well.

Symptoms

Red thread is easily recognized by the red to coral pink fungal strands (stroma) produced on leaf blades and sheaths. Early in the morning when the dew is still present, or on rainy days, the fungal stroma looks gelatinous and fleshy. As the grass dries, the stroma also dries and becomes thin and thread-like. When seen from a distance, infected turf often appears to be suffering from lack of water; from a little closer, it looks as though it has leaf spot; and from directly above, especially in red fescue, the disease resembles dollar spot. A close examination of the grass blades, however, reveals red to coral-pink stromata. Under conditions of high humidity, pink cottony flocks of arthroconidia can be seen in the patches. A red thread infection of a fine-leaf fescue turf is pictured in Figure 3.5.

Occurrence

Laetisaria fuciformis survives as stromata and dormant mycelium in host tissue. Development of the disease is favored by high humidity at temperatures between 60–90°F (15.5 to 32°C). The cool drizzly days of spring and fall are ideal for the development of red thread, and the disease spreads from plant to plant by the growth of the stroma. When they break up, the stromata can be carried long distances by the wind. They are often cut off by mowing, and then spread by wind and mowing equipment.

Figure 3.5. Red thread in a fine-leaf fescue turf.

MANAGEMENT

Cultural

The obvious cultural practice is fertilization with nitrogen. This is a good solution in the summer. In the spring and fall, the answer is not so simple. The areas of the cool-season grassbelt in which red thread is most severe are the same areas in which *Microdochium* patch is most serious, and the nitrogen treatment that alleviates the red thread problem will aggravate *Microdochium* patch. In addition, *Typhula* blight and desiccation can be a problem in these areas and increasing the nitrogen level will make the turf more succulent and consequently more susceptible to *Typhula* blight and desiccation.

What should be done? Since red thread is neither as severe a disease as *Microdochium* patch or *Typhula* blight nor as destructive as desiccation, a good nitrogen fertilization program should be followed in the summer. The date of the last application will vary, but it should be early enough to allow the turf to harden off before the permanent snowfall. Then an appropriate fungicide can be used to manage red thread. Fortunately, there are many fungicides that are effective.

Genetic

No conclusive data are available yet on cultivars resistant to red thread. Because it is a minor disease that can be managed relatively easily by cultural practices and fungicides, selecting a cultivar solely for its resistance to red thread is generally not practical. Cultivars should be chosen for their resistance to the turfgrass disease that is the most serious threat, and in most areas that is not red thread. Although turfgrass cultivars that are resistant to red thread and *Microdochium* patch might be very important in the Pacific Northwest, New Zealand, and in parts of Europe, they would be of minor importance in most other areas of the world. As mentioned above, data on varietal reaction to red thread are inconclusive. However, perennial ryegrass should not be grown in localities where red thread is a problem. The improved perennial ryegrasses are another example of cultivars chosen for an agronomic characteristic (mowability) with little or no regard for potential disease problems. In northern coastal regions of the United States, Europe,and New Zealand, red thread is a serious disease of the perennial ryegrasses and colonial bentgrass, requiring fungicide treatments during the cool wet weather of the fall.

Chemical

High-maintenance turf, especially the perennial ryegrasses, may require some type of chemical management in the fall and winter. This is necessary because late-season nitrogen applications that might alleviate the red thread problem will increase the severity of the more serious diseases *Microdochium* patch and *Typhula* blight. Fall nitrogen applications that cause the turf to be lush before the advent of freezing weather can also cause serious turf loss from desiccation during open winters. In cool climates, several weeks of cool fall weather, especially with cold rain, can lower the soil temperature to a point where grass growth is minimal even when daytime temperatures are adequate for growth. Under these conditions, sufficient or even excess nitrogen does little to stimulate top growth, and fungicide treatments are necessary to prevent severe turf loss.

DISEASE: Fairy rings

Pathogen: Many fungi, all in the class *Basidiomycetes*

Hosts: All turfgrasses

Fairy rings are caused by many soil-inhabiting fungi of the class *Basidiomycetes*. For a list of the different species associated with this disease, see Reference #137. For identification of the specific fungus causing a particular ring, you will need to consult an expert who is experienced in identifying *basidiomycetes* fungi. In general, identification of the species involved is not really necessary if control is your prime concern. If fairy rings appear, you can do one of three things: fumigate, excavate, or try to enjoy them for their aesthetic beauty. When fairy rings occur on greens, however, the third solution is not recommended for golf course superintendents who want to keep their jobs.

Symptoms

Fairy ring fungi tend to grow in a circle through the organic matter in the soil, mat and thatch, breaking down organic matter as they grow and releasing nitrogen (ammonia) that eventually is reduced by other soil microorganisms to nitrates. This nitrogen stimulates turfgrass growth ahead of the dead or dying turf and sometimes behind it as well (see Figure 3.6). The area of increased growth—known as the zone of stimulation or the activity zone—is where the turf may die if heat or

Figure 3.6. Fairy ring on a bowling green.

drought stress occurs. If the turf is watered it will usually recover, although it may still be stunted.

What causes the yellowing, stunting, or dying in the zone of activity? The yellowing and stunting certainly suggest some type of parasitic activity. Filer[53] demonstrated penetration and death of the cortical cells by the fairy ring-producing fungus *Marasmius oreades*. Filer[53,54] also suggests that hydrogen cyanide production by *M. oreades* is a possible explanation for the pathogenicity of the fairy ring fungi. The presence of the mycelium of the fungus in the thatch and soil could also account for the death of the grass plant. The mycelium is hydrophobic and consequently is impervious to water, creating a desert below it. Once the soil under the fungal mycelium becomes dry, it is difficult to wet again by artificial irrigation unless a wetting agent is added. A cross-sectional diagram of a fairy ring in a turfgrass profile is shown in Figure 3.7.

Occurrence

By and large, fairy rings are caused by fungi that live or survive by spores or other means in dead organic matter in the soil or thatch. Areas that previously were forested, or ones in which stumps or logs have been used as fill, are prime candidates for the development of fairy rings. Fairy rings are also very prevalent in high sand content soils, probably due to the lack of competition from other microorganisms.

It is still unclear what causes the regrowth of dormant mycelium or initiates spore germination, but it is undoubtedly related to the phenomenon of fungistasis. Beard et al.[7] showed that 3 or more pounds of nitrogen or nitrogen sources might be important, since the fairy ring development started after the nitrogen source was changed from ammonium nitrate to urea. Pesticides may also cause the development

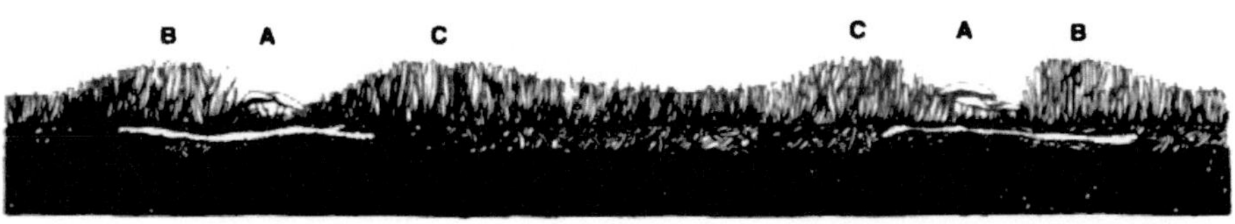

Figure 3.7. Cross section of a fairy ring in a turfgrass profile. The structure is typical of a *Tricholoma*-type fairy ring, which very seldom causes death of the turf. The mycelium (main body) of the fungus usually forms a thin layer that is confined to the thatch. (*A*), stunted turfgrass with mycelium evident in the thatch; (*B*), outer zone of stimulation; (*C*), inner zone of stimulation.

of fairy rings by destroying saprophytic bacteria and fungi that normally remove exogenous nutrient sources. The nutrients that build up can serve as food for germination and growth of fairy ring fungi.

MANAGEMENT

Management or removal of fairy rings is difficult and expensive, and you should study the total situation carefully before making a decision. If the fairy rings are on a putting green they definitely should be removed. But if they are in the fairway they should generally be ignored, with the possible exception of a large concentration in a critical land zone or the approach to a green, in which case the entire area should be fumigated and reseeded. Fairy rings should be removed by fumigation from sod fields for obvious reasons. On areas used for athletics, on commercial areas, or on home lawns they should be ignored or turned into a conversation piece, unless they are numerous or really unsightly. If almost the entire area is covered with them, or if they are bothering you, have the sod removed and the entire area fumigated. If you have only a few small ones, you may wish to dig them out.

Cultural

The cultural practice most often recommended for management is fertilization with nitrogen to stimulate growth of the surrounding grass so that it will mask the tall dark green grass in the zone of stimulation. This is where identification of the fairy ring fungus becomes important. Adding nitrogen to mask fairy ring symptoms caused by *Tricholoma* species may only result in more fairy rings,[7] and the same is probably true for many of the other species of fungi that cause fairy rings.

Another method consists of trying to drown the fairy ring by supersaturating the area with water for 48 or more hours. This treatment has not been effective in all cases and the fairy rings always return later that season or the next. In other words, you can't drown them.

The best way to manage fairy rings is to dig them out. Go a foot (.3 meter) beyond the ring, square it off, remove the sod, and then remove the soil within the square to a depth of 1 foot (.3 meter). Fill the hole with uninfested soil, and reseed or resod (with new, and hopefully, fairy ring-free sod).

Genetic

Since fairy ring fungi are basically saprophytic, the concept of resistant cultivars is not applicable.

Chemical

You must learn to like fairy rings or else fumigate them or dig them out. The information about drilling holes and pouring fungicides down them to eradicate fairy rings isn't worth the paper it is written on. The result of such a management procedure is usually replacement of the original fairy ring with several new ones, which develop from the unkilled portion of the first ring. If you have put on a few pounds over the winter this may be good exercise, but it won't get rid of your fairy rings.

DISEASE: Gray leaf spot

Pathogen: *Pyricularia grisea* (Cke.) Sacc.

Host: St. Augustinegrass, perennial ryegrass, bermudagrass, crabgrass

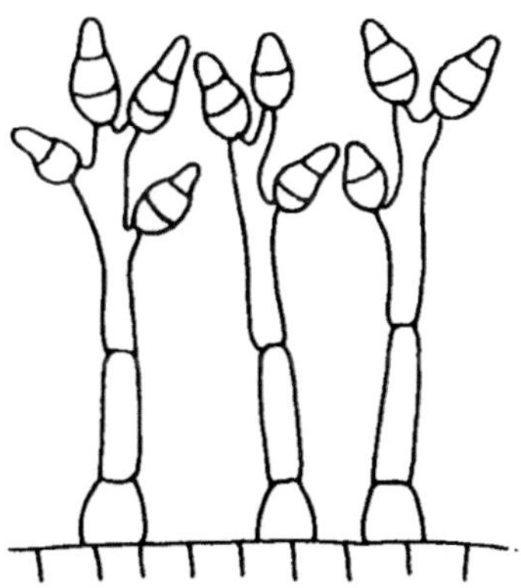

Conidiophore and condiospores of *Piricularia grisea*

Gray leaf spot is considered a problem primarily on St. Augustinegrass, although it has been found on bermudagrass, crabgrass, foxtailgrass, and barnyardgrass.[146] However, recently it has been reported as a problem on perennial ryegrass, particularly young seedlings.[88] The disease is most severe during warm, humid weather.[58] On St Augustinegrass it occurs primarily along coastal areas in the southern U.S. It occurs on perennial ryegrass wherever it is grown.

Symptoms

The leaf lesions first appear as tiny brown spots that enlarge and become oval or elongated. The mature spots usually have depressed gray centers with irregular brown margins surrounded by a ring of chlorotic tissue. If foliar infection is severe, the leaves may appear burned or scorched. Lesions also are found on the leaf sheath, spike, and stems. Diseased areas on the leaf sheaths and spikes are similar to those found on the leaf blades, whereas those found on the stem are often brown to black. These symptoms are illustrated in Figure 3.8.

Occurrence

Pyricularia grisea overwinters as spores and dormant mycelium in the lower leaves of infected plants and in the thatch. As the temperature rises, the spores are blown by the wind and splashed by water to healthy plants which they can infect. Under conditions of high humidity and high temperature (80° to 90°F [26° to 32°C]), the disease may develop rapidly.[58] Newly sprigged, rapidly growing grass is more susceptible to gray leaf spot than are well-established St. Augustinegrass turfs.[146]

Figure 3.8. Close-up of foliar symptoms of gray leaf spot. (Photograph courtesy of Dr. Bobby G. Joyner.)

MANAGEMENT

Cultural

Avoid excessive nitrogen levels or applying nitrogen just before and during the time of year when gray leaf spot is a problem. Irrigation in the middle of the day will allow foliage to dry quickly and thus help reduce the severity of the disease.

Genetic

The purple-stem types of St. Augustinegrass, such as Roselawn and Floratam, are more resistant than the green-stem types, such as Bitter Blue and Floratine (T.E. Freeman, personal communication).

Chemical

Gray leaf spot can be managed by weekly applications of one of the fungicides recommended in Chapter 6 and in the Appendix.

DISEASE: Slime mold

Pathogen: *Mucilago crustacea* (Leyss.) Morg., *Physarum cinereum* (Batsch) Pers.

Hosts: All grasses

Slime molds are not pathogenic on turf, although they may strike fear into the heart of the novice turfgrass manager. These saprophytes usually do nothing more than cause an unsightly appearance. If they persist for several days on the foliage they may cause chlorosis, but serious permanent damage seldom occurs.

Physarum cinereum is more common than *Mucilago crustacea*.[18] The fruiting bodies can be observed on the foliage of the grass, on clover and other weeds, and on the soil or thatch. They consist of grayish-white fruiting structures filled with purple spores, and often resemble cigarette ashes. Spots in the turf are usually irregular circles a few inches (5 cm) to a few feet (1 meter) across. If wet, cool weather persists, large irregular areas may become covered with slime mold. A typical slime mold infestation is illustrated in Figure 3.9.

Figure 3.9. Slime mold on turf and weed foliage. (Photograph courtesy of Dr. Don Blasingame.)

MANAGEMENT

Chemical management is not necessary; in fact, it is a waste of time and money. If the infestation is heavy, poling, brushing, or a sharp spray of water from a hose will remove the slime mold from the leaves.

CHAPTER 4

Fungal Diseases of Turfgrass III: Disease Complexes

THE PATCH DISEASES

There is now a group of diseases caused by several fungi that attack the root systems of the various turfgrasses. They all produce ectotrophic hyphae called "runner hyphae" on grass roots. These are relatively new diseases, or at least the identification of the correct causal organisms has occurred over the past 10 years. For an excellent review of these diseases, read Reference #21.

DISEASE: *Fusarium* blight. Extinct disease. Please see summer patch and necrotic ring spot.

DISEASE: Bermudagrass decline

Pathogen: *Gaeumannomyces graminis* (Sacc.) Arx & D. Olivier var *graminis*

Host: Bermudagrass

Bermudagrass decline is a relatively new disease, at least from the point of recognition. It has been referred to by many golf course superintendents as the bermudagrass simply "getting old and dying." When you try to explain to them that it is a disease, the typical reaction is, "No, it is not a disease. Diseases only occur on overseeded or northern grasses. This is just bermudagrass getting old." Another bit of folklore that has lived too long. It's like the one that says annual bluegrass dies from the heat and not from diseases. Hopefully, work done by Elliott[43] will retire this old piece of folklore to the turfgrass museum of useless relics.

Symptoms

The symptoms first appear as yellow patches ranging in size from 6 inches (15 cm) to 3 feet (1 meter) in diameter. The patches eventually turn dark brown in color, and may eventually become devoid of turf. The roots on the infected plants are dark-colored, and may have dark strands of fungal mycelium (runner hyphae).

Occurrence

The disease occurs in late summer through the fall in the southeastern United States. It has been associated with heavy rainfall during this period, as is the case with necrotic ring spot and summer patch.

Bermudagrass decline has primarily been reported on bermudagrass greens, but *G. graminis* has been observed on the roots of higher cut bermudagrass collars, tees, and fairways. Perhaps what we are dealing with is a fungus that can cause harm to shorter-cut bermudagrass that is under more stress during the growing season. The higher-cut bermudagrass can withstand the disease during the growing season. The symptoms observed on the higher-cut bermudagrass may occur during the dormant period when cold weather causes these weakened infected plants to die over the winter. The disease we see in the spring is what we call spring dead spot. What I am suggesting is that we are dealing with two phases of the same disease: one that occurs on low-cut bermudagrass in the late summer and early fall, and the other called spring dead spot that occurs on higher-cut bermudagrass during the dormant period. It becomes even more likely when you consider the same fungus, *G. graminis*, is responsible for both diseases in the southeastern United States.

MANAGEMENT

Cultural

Little is known about the management of bermudagrass decline. One can only extrapolate from the other patch diseases, and this is always dangerous. However, until research can gives us the data we need, educated guessing will have to do.

The severity of most of the other patch diseases is reduced when moderate amounts of nitrogen are applied as a complete fertilizer, with the slow release forms of nitrogen being more effective than the quick release.

Light daily irrigation has also been shown to reduce the severity of the other patch diseases, and deep irrigation to increase the severity.

Genetic

All improved cultivars of bermudagrass are susceptible, as well as common types.

Chemical

Once again, little is known about the chemical control of bermudagrass decline, but based on the other patch diseases like summer patch, where the symptoms also occur in the late summer and early fall, fungicides should be applied beginning in the spring period. Because of the problems associated with the DMI fungicides on bermudagrass, you should check with your local turfgrass specialist for specific recommendations.

DISEASE: Necrotic ring spot

Pathogen: *Leptosphaeria korrae* J. C. Walker & A. M. Sm.

Host: Kentucky bluegrass, fine-leaf fescue, annual bluegrass, rough bluegrass

Occurrence

Necrotic ring spot, caused by *Leptosphaeria korrae*, is the most commonly occurring patch disease on Kentucky bluegrass turfs (see Figure 4.1). For many years this disease was also called *Fusarium* blight, but critical research[15,124,164] has shown the true cause of this disease to be *L.*

Figure 4.1. Necrotic ring spot in Kentucky bluegrass lawn.

korrae, and not species of *Fusaria*. Once the real cause of the disease was known, developing an integrated management program for necrotic ring spot became much easier.

Symptoms

The disease first appears as small patches six inches (15 cm) to a foot (.3 meter) in diameter. When *L. korrae* is actively attacking the Kentucky bluegrass plants in the fall of the year, dark red blades of grass can be seen in the patches. In the summer, the infected plants in the patches, where root systems were destroyed by the pathogen in the spring and fall, begin to wilt even though the turf in the surrounding area is healthy. The infected plants in the patches eventually turn straw-colored. Patches that are two years old or older have what is referred to as a "frog eye" appearance. This symptom, described as having a center of healthy turf surrounded by a ring of dead turf, is caused by *L. korrae* moving out of the center of the ring to the outer edges. The turf that wasn't killed in the center of the patch begins to recover where the fungus is no longer active.

MANAGEMENT

Necrotic ring spot is the one turfgrass disease for which a complete IPM program has been developed. The IPM system for managing necrotic ring spot includes all four means of managing plant diseases: biological, cultural, genetic, and chemical.

Biological

The pathogen is active in the cool weather of the spring and fall, even though the symptoms are seen in the warm weather of the summer. The summer symptoms are the result of the turfgrass not being able to take up adequate amounts of water during the summer stress period because the fungus had destroyed the root system during the cool weather. What we are trying to do is keep these weakened plants alive during the warm weather of the summer. The best way to accomplish this is through the use of light, daily irrigation. The lawn should be irrigated every day with 0.1 inch (.25 cm) of water, preferably applied between 12:00 noon and 4:00 P.M. During prolonged droughts when measurable rain does not occur during a two-week period, the irrigation rate should be raised to 0.2 inch (.5 cm) per day. The daily irrigation regime also keeps the thatch and soil more moist than a conventional deep, infrequent irrigation regime. Having moist soil and thatch results in high populations of bacteria being maintained. These bacteria have been shown to produce antifungal compounds that are inhibitory to the growth of *L. korrae* in laboratory culture studies. These bacteria are believed to aid in the prevention of new infections in the late summer and early fall.

The light, daily irrigation not only supplies adequate water during the stress period of the day, but also cools off the plants to allow them to get through the stress period. Research has shown that summer symptoms can be completely eliminated through the use of light daily irrigation.[155]

Lawn Restore and Turf Restore have been shown to biologically manage this disease. The products contain beneficial organisms as well as a food source to stimulate the growth of the natural microflora. These organisms have also been shown to inhibit the growth of *L. korrae* in laboratory antagonism studies.

Cultural Practices

Successful management of necrotic ring spot requires adequate levels of nitrogen. It obviously requires adequate levels of phosphorus and potassium as well, but the soil can be tested to determine the level of these nutrients, and they can be added accordingly to bring soil levels of these nutrients up to acceptable levels. Nitrogen, on the other hand, is not held by the soil and must be applied every year.

Improved Kentucky bluegrass cultivars require between 4 to 6 lb (2 to 2.9 kg) of actual nitrogen per 1000 sq ft (100 square meters) per season. The slow release forms of nitrogen contained in fertilizers like Turf Restore and Lawn Restore, IBDU, UF, etc., are also more effective in managing the disease than fast release fertilizers like urea. Complete fertilizers containing phosphorus and potassium as well as slow release nitrogen carriers are the most effective. Daily irrigation, applied from midday through 4:00 P.M. will help cool off the turfgrass plants. This allows the infected plants with depleted root systems to survive the heat stress of the day.

Chemical

Curative

The benzimidazole fungicides, benomyl and thiophanate methyl, work best. This is because they have cytokinin-like properties which tend to stimulate growth. To be effective, these fungicides need to be drenched into the soil before they dry on the foliage, since they are only translocated upward, and this is a root disease. Best results are obtained when the turf is irrigated before the fungicides are applied.

Preventive

The DMI fungicides like fenarimol and propiconazole or the dicarboximide fungicide iprodione work best preventively. Part of the reason they do not work well curatively is that fenarimol and propiconazole have plant growth regulator properties, which, when used at high rates to control necrotic ring spot, slow down the growth of the plant. Even if they stop further infection by the fungus, recovery is not evident because of the slowed growth of the plants. These fungicides are translocated in both directions, and therefore do not have to be drenched into the soil to be effective.

Our studies have clearly shown that unless adequate nitrogen is applied, these fungicides will not be effective.

Genetic

The Kentucky bluegrass cultivars that have been resistant in inoculated field trials were Midnight, Monopoly, Able 1, Majestic, and America.

DISEASE: Spring dead spot

Pathogen: *Leptosphaeria korrae*, J. C. Walker and A. M. Smith,
Leptosphaeria narmari, J. C. Walker and A. M. Smith
Gaeumannomyces graminis (Sacc.) Arx & Olivier,
Ophiosphaerella herpotricha (Fr.) Sacc.

Host: Bermudagrass

Spring dead spot (SDS) causes more damage to bermudagrass, particularly in the cooler parts of the warm-season grass regions, than any other single disease. It is discouraging to have beautiful lush green bermudagrass fairways, greens, and home lawns in the fall, before the bermudagrass becomes dormant, only to find large areas that fail to green up in the spring because they have been destroyed by spring dead spot.

Symptoms

Spring dead spot is first evident in the spring after the bermudagrass breaks dormancy and begins to green up. It appears as brown, more or less circular spots in the turf, ranging in diameter from a few inches (5 cm) to 3–4 feet (.9 to 1.2 meters) (Figure 4.2). Except for the well-defined circular spots so typical of fungal diseases, the grass looks as though it has simply failed to break dormancy; the foliage has the typical straw color of dormant bermudagrass. Examination of the stolons and roots, however, reveals black discoloration. In late spring the dead spots may have green centers, which are often the result of invasion by broadleaf weeds and weed grass, especially in the northern regions of the bermudagrass belt. In the warmer portions of the bermudagrass belt, where good broadleaf weed management is practiced, the spots will be covered over by bermudagrass stolons growing in from the periphery of the spots.

Figure 4.2. Spring dead spot in a home lawn turf. (Photograph courtesy of Bill Small.)

Occurrence

For many years the cause of spring dead spot remained unknown, even though many fungi, including *Helminthosporiums, Curvularias, Pythiums, Fusaria*, and *Ophiobolus graminis*, have been isolated from disease tissue. All attempts to reinoculate and produce spring dead spot symptoms failed. Finally researchers[19,51,97,145,165,166] began to isolate organisms with which Koch's postulates could be completed. It now appears that a group of fungi, all of which attack the root system and produce ectotrophic hyphae, attack the bermudagrass turf during the growing season, weakening the plants, which are then killed by the cold weather stress of winter. This is similar to what happens to annual bluegrass, which is attacked by *Magnaporthe poae* in the spring but doesn't die until the summer stress occurs.

It should follow, then, that the severity of spring dead spot is dependent on the severity and length of winter dormancy. The longer the period of dormancy and the colder the temperatures, the more severe the disease. Bermudagrass is known to be susceptible to cold-temperature kill, so it only stands to reason that weakened bermudagrass is even more susceptible. Therefore, spring dead spot is most severe in the northern part of the bermudagrass region where the winters are also more severe.

MANAGEMENT

Cultural

Cultural practices that reduce thatch, such as vertical mowing and topdressing, also help manage spring dead spot, presumably because they lessen the amount of thatch. Lucas[96] showed that extra applications of nitrogen during August and September made spring dead spot worse and resulted in lower-quality turf the following spring. Dernoeden et al.[34] showed that ammonium-based fertilizers in combination with potassium helped reduce SDS over time.

Genetic

At this time there do not appear to be any truly resistant cultivars. The improved cultivars like Tiffine, Tifgreen, and Sunturf are reported to be more susceptible than common bermudagrass, but their "susceptibility" may simply be due to the fact that they usually receive more maintenance (i.e., more nitrogen) than common bermudagrass does.

Chemical

Wilcoxen's work[168] has shown that fungicides applied to bermudagrass turfs during the growing season will reduce the severity of spring dead spot. The fungicide used in the study was chlorothalonil, which was applied biweekly at a concentration of 3 ounces per 1000 square feet (95.4 mL per 100 square meters). One test area received treatment during April and May, a second area was treated during June and July, a third during August and September, and a fourth from April through September. There was, of course, an untreated control plot. The April-May area received only three treatments because the study was not initiated until late April. When data were collected from the plots the following spring, all the treatments had either reduced spring dead spot or maintained it at the same level. The untreated control plot, on the other hand, had too many spots even to be read. Sturgeon and Jackson,[140] working with large amounts of PCNB + ethazole applied as a drench, showed that a single application may provide control for more than one season. Lucas[96] managed spring dead spot in the spring of 1974 with six monthly applications of benomyl from July through

December 1973, in an area that had severe spring dead spot in the spring of 1973. Lucas[96] also investigated the effect on disease management of the timing of fungicide application. He was able to prevent the occurrence of spring dead spot by applying benomyl at monthly intervals during the fall (October, November, December), but he found that applications of benomyl at other times of the year did not control the disease.

The cost of treating golf course greens is certainly within most budgets. Whether or not such a program is economical for fairways will have to be determined on a course to course basis. You should keep in mind that $7,000 or $8,000 a year for a fungicide program is cheaper than resprigging the course every 4 to 5 years, not to mention the inconvenience to the golfers caused by the reestablishment operation or by not reestablishing and playing on weeds.

DISEASE: Summer patch

Pathogen: *Magnaporthe poae*, Landschoot & Jackson

Host: Annual bluegrass, Kentucky bluegrass, fine-leaf fescue

For many years, it was believed that annual bluegrass died from high temperatures. Research has since shown that annual bluegrass actually dies from three major pests in the warm weather: anthracnose, the black turfgrass ataenius beetle, and summer patch. The most devastating and most difficult to control of all of these is summer patch, caused by *Magnaporthe poae*. Summer patch is also a serious disease of Kentucky bluegrass grown in areas of the world that have very warm summers, where the nighttime temperatures remain above 70°F (21.1°C) for a month or more (see Figure 4.3).

Originally the disease was mistakenly diagnosed as *Fusarium* blight. Later it was called summer patch by Smiley and Craven-Fowler,[132] who isolated the causal fungus and identified it as *Phialophora graminicola*, and completed Koch's postulates. Landschoot and Jackson,[87] upon further study, correctly identified the fungus that causes summer patch as *Magnaporthe poae*.

Symptoms

The disease first appears in the warm weather of summer as yellow to bronze-colored, irregular-shaped patches ranging in size from 6 inches (15.3 cm) to 3 feet (.9 meter) in diameter. The patches may coalesce,

Figure 4.3. Kentucky bluegrass turf infected with summer patch.

resulting in areas of affected turf that are several feet in diameter. In annual bluegrass-creeping bentgrass polystands, the annual bluegrass is damaged and the creeping bentgrass is not affected, often resulting in a "frog-eye" symptom. Examination of the roots with a dissecting scope will reveal dark-colored ectotrophic hyphae called runner hyphae.

Occurrence

The disease first appears in the warm weather of the summer, usually following a rainy period or heavy downpour. Heavy or excess irrigation during this period will also bring on symptoms. Modeling studies have shown a direct correlation between moisture pushing the oxygen out of the soil and the development of symptoms. Kackley et al.[81,82] also showed symptoms to be most severe where moisture was not limiting. The other key factor for symptom development is a soil temperature above 70°F (21.1°C) at a 2-inch depth for at least 48 hr. Even though symptoms occur during the warm weather, the initial infection occurs in the spring of the year when the soil temperatures first reach 65°F (18.3°C) at a 2-inch soil depth. Only the outer cortical tissue of the root is infected at this time. *M. poae* remains in the cortical tissue, causing little noticeable damage, until the soil temperatures reach above 70°F

(21.1°C) and the rain or deep irrigation regimes push the oxygen out of the soil. This weakens the roots of the grass plant, and allows the pathogen to invade the vascular tissue of the plant, where it interferes with water and nutrient uptake and movement of photosynthesis products from the foliage to the roots.

MANAGEMENT

Cultural Practices

Fertility

Appropriate fertility starts with having adequate levels of all of the nutrients present in the soil. Most of these can be applied based on soil tests. Nitrogen, of course, is the exception. Having adequate levels of nitrogen is not only important in managing summer patch, it is essential. Without adequate levels of nitrogen, even fungicide treatments will not be effective. Research has clearly shown that to manage summer patch effectively, a half pound of nitrogen per 1000 sq ft (.25 kg per 100 square meters) every three weeks is required. The slow release carriers, like Turf Restore and IBDU, have been more effective in managing summer patch than quick release forms of nitrogen, like urea. In spite of "folklore" which suggests nitrogen should not be applied in the summer because it will make the plant more susceptible to stress, research shows that moderate levels of nitrogen are essential in order to manage summer patch.[38]

Irrigation

Light daily irrigation ranging in amounts from 0.1 to 0.4 inches (.25 to 1 cm), depending on rainfall, temperature, and evapotranspiration, will help reduce the severity of summer patch. The most effective time to apply this irrigation water is mid- to late afternoon. This allows time for the foliage to dry before entering the nighttime period. It also helps to cool the plant off during the warmest part of the day to better allow it to survive summer stress. This type of irrigation regime does not push the oxygen out of the soil, so as to bring on summer patch symptoms. Of course nothing can be done, short of building a tent over the golf course, to prevent heavy rains from falling, but at least the light daily irrigation will not add to the problem and may delay the onset of a summer patch epidemic.

What about the old theory that light daily irrigation will result in the turfgrass plant having a short root system? Unfortunately, that's just what it is, a theory. I challenge you to find research data to support the claim. It is based on "folklore" rather than good scientific data. Rhizotron and other studies have shown that soil temperatures during the warm summer months determine the length of the root system. This *light daily* irrigation is done *every day* as a preventive treatment, and is not to be confused with a study where symptoms were allowed to develop and then light infrequent irrigation was applied.[38] Also, heavy irrigation that replaces oxygen in the soil will make the disease more severe.

Chemical

Preventive

For many years trying to manage summer patch with fungicides proved to be erratic at best and futile the rest of the time. The key turned out to be timing. The best fungicides for the management of summer patch are the DMI fungicides, triadimefon, fenarimol, and propiconazole; but to be effective they have to be applied before symptom development occurs. Triadimefon and fenarimol need to be applied as soon as the soil temperatures reach 65°F (18.3°C), with a second application being made 30 days later. In areas where summer lasts for more than 3 months, an additional application may have to be made 30 days later. The first application of propiconazole should be made when the soil temperatures reach 75°F (23.4°C) at a 2-inch depth, with a second application being made 30 days later.

Curative

Once the disease symptoms are present, the benzimidazole fungicides, like benomyl and thiophanate methyl, manage the disease the best; however, some loss of turf is encountered when trying to manage summer patch curatively. A second application three weeks later is usually required with these fungicides. Remember, to be effective these fungicides need to be drenched into the root zone.

Unless adequate levels of nitrogen are also applied, fungicide treatments will not be effective.

DISEASE: Take-all patch (formerly called Ophiobolus patch)

Pathogen: *Gaeumannomyces graminis* (Sacc.) Arx & Oliv. var. *avenae* (E. M. Turner) Dennis

Hosts: Colonial bentgrass, creeping bentgrass, and velvet bentgrass

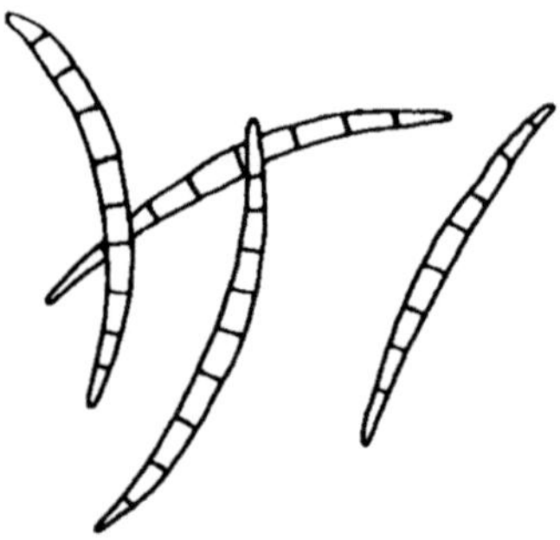

Gaeumannomyces graminis conidiospores

Take-all patch is a very serious disease, especially on newly established creeping bentgrass turfs. It occurs in England, northern Europe, South Africa, New Zealand, Australia, and throughout North America. It has even been reported in the deserts of the southwestern U.S. where permanent bentgrass greens are maintained.

Symptoms

Take-all patch appears at first as a patch of bronzed or bleached turf. The grass in the center eventually dies and is replaced by broadleaf weeds, and more often, by annual bluegrass (Figure 4.4). At first, the spots are 4 to 6 inches (10 to 15 cm) in diameter, but they may grow to a diameter of several feet (1 meter) over a period of years. The plants in these patches have poorly developed roots and are easily pulled out. The black runner hyphae of the fungus can be seen under the base of the leaf sheath, crowns, and on the roots.

Occurrence

Gaeumannomyces graminis can survive as a saprophyte in the thatch and as dormant mycelium in the perennial parts of the grass plant. It is a fungus that attacks primarily the crown and root of the plant, usually during the cool, wet weather of the spring and fall. It is most serious on newly established creeping bentgrass turfs, usually reaching its peak in 3 to 5 years and then gradually declining. This is believed to be due to the natural buildup of microorganisms antagonistic to *G. graminis* in the soil.

According to Smith et al.,[137] the disease is favored by applications of

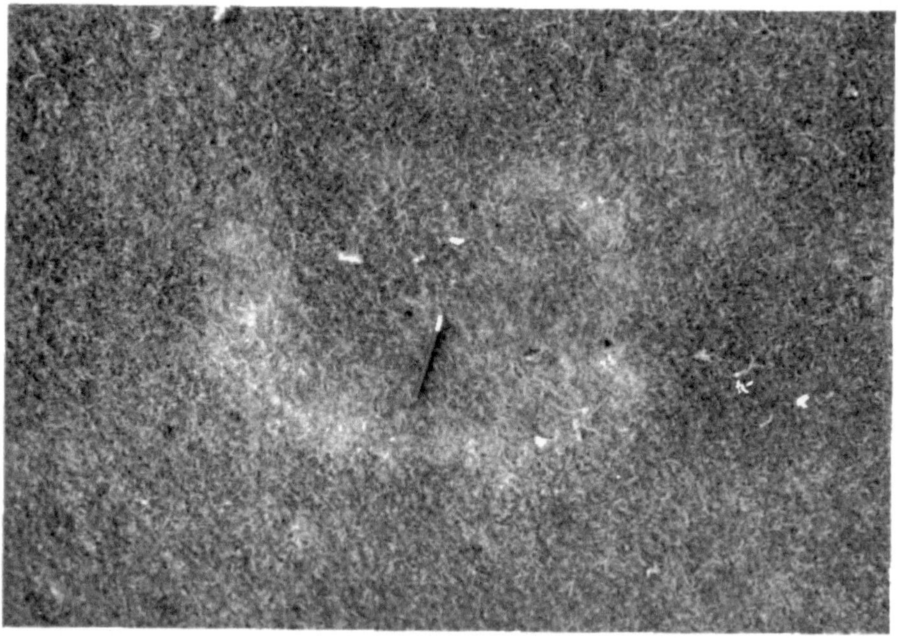

Figure 4.4. Take-all patch on a young creeping bentgrass green.

lime and by high pH. Dernoeden[32] showed a decline in the amount of take-all patch when acidifying fertilizers like ammonium sulfate and ammonium chloride were used.

MANAGEMENT

Cultural Practices

Adequate levels of phosphorus and potassium should be maintained. Liming should be avoided in areas where take-all patch is a problem. The problem with using fertilizers like ammonium sulfate and ammonium chloride is that they have a high salt index and can injure the turf. Even where severe injury from these fertilizers does not occur, a thinning of the turf is observed compared to low salt index fertilizers.

Sulfur[62] will help to manage the disease. The problem with using sulfur to manage take-all patch is that it can lead to the black layer (see Chapter 5). It is like trying to cure the common cold with a drug that will give you cancer. It is much healthier for your job to manage the disease with fungicides.

Genetic

Creeping bentgrass appears to be the species of turfgrass most susceptible to take-all patch. There are no reported resistant cultivars at this time.

Chemical

A good timing model needs to be developed for optimum disease management with fungicides. At the current time the best recommendation is to apply fenarimol or propiconazole after the first mowing, followed by a second application 30 days later. This needs to be repeated in the late summer following the first night when temperatures drop into the 50s F (50°F = 10°C).

THE SNOW MOLDS

DISEASE: *Typhula* blight (gray snow mold)

Pathogen: *Typhula incarnata* Lasch ex Fr., (syn. *Typhula itoana* Imai), *Typhula ishikariensis* Imai

Hosts: Annual bluegrass, colonial bentgrass, creeping bentgrass, fine-leaf fescues, Kentucky bluegrass, perennial ryegrass, tall fescue, velvet bentgrass

Typhula blight is an important disease in the northern areas of the Northern Hemisphere where snow stays on the ground for extended periods. It is most important in regions where snow cover remains on the ground for three months or more without a thaw. It may be found in combination with *Microdochium* patch (pink snow mold), but its range does not extend as far south as that of *Microdochium* patch. Unlike *Microdochium* patch, which occurs with or without snow cover, it does require some type of cover. The cover is usually provided by snow, but leaves, straw mulch, and desiccation covers will also suffice.

Typhula incarnata is the most significant pathogen producing gray snow mold in the eastern United States and Canada, whereas *Typhula ishikariensis* prevails from Michigan west. *T. ishikariensis* is more prevalent in areas where the snow cover remains for four or more months. *T. ishikariensis* was reported by Vaartnous and Elliott to be the most important pathogen causing snow mold on red fescue and Kentucky bluegrass turfs in northwestern Canada in 1968.[151] J.D. Smith[136] reported widespread occurrence of the fungus in western Canada, and Stienstra[139] has reported that *T. ishikariensis* is an important pathogen in Minnesota.

Symptoms

At temperatures between 30° and 55°F (-1 and 12.7°C), the *Typhula* fungus grows and infects. As the snow melts, circular grayish or straw-colored to dark brown infection centers appear in the turf. The spots range from 3 inches (7.6 cm) to 2 feet (0.6 meter) in diameter, but most are between 6 and 12 inches (15 and 30 cm) across (Figure 4.5). Immediately after the snow melts, the grayish white mycelium of the fungus can be seen, especially at the outer margins of the spots. The disease gets its common name, gray snow mold, from the color of the mycelium. *Typhula* blight is worse in winters when the snow falls on unfrozen turf, or on turf that has not been hardened off by frost. Often when snow falls on frozen ground, the disease develops in the spring when the snow begins to melt.

The two species can be easily separated when viewed shortly after

Figure 4.5. Disease free fungicide treated golf course green compared to a creeping bentgrass fairway severely infected with *Typhula* blight. Both green and fairway are Penncross creeping bentgrass, while unaffected grass around the bunker is Kentucky bluegrass.

snow melt. *T. incarnata* produces large rust-colored sclerotia in infected turf, whereas *T. ishikariensis* produces small black-colored sclerotia, which appear as though someone had sprinkled the turf with pepper.

Occurrence

The life cycle of *T. incarnata* is illustrated in Figure 4.6. The *Typhula* fungi oversummer as sclerotia. In early spring, when the snow is melting, the sclerotia may be as large as 3/16 of an inch (.48 cm), and can easily be seen with the naked eye. Later on, they dry and shrink and become invisible (or at least hard to detect). As sclerotia, the fungus can survive unfavorable warm temperatures and the fungicides that may be part of a summer program of disease control. With the arrival of cool wet fall weather and in the presence of ultraviolet light, the sclerotia swell and germinate, producing pink, club-shaped

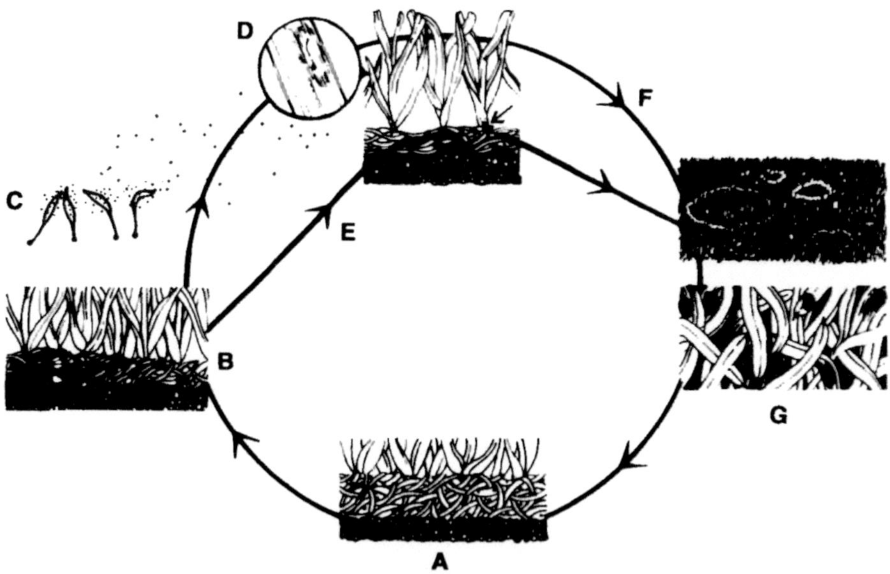

Figure 4.6. Life cycle of *Typhula* blight caused by *T. incarnata*. (A), *Typhula* oversummers as sclerotia in the thatch or mat; (B), sclerotia swell with the advent of cool weather and germinate, producing basidiocarps when exposed to ultraviolet light; (C), basidiocarps produce many basidiospores, which are disseminated by the wind; (D), basidiospores germinate and infect turfgrass; (E), some sclerotia germinate and infect turfgrass directly, without production of basidiocarps; (F), *Typhula* infects the turf under snow cover or in the melting margins of the snow during the spring thaw; (G), close examination of the turf immediately after the snow has melted reveals the presence of light brown to reddish brown sclerotia.

sphorophores,[115] from which basidia and basidiospores arise. The basidiospores are carried by the wind to new places or to areas where the disease was eradicated last season. When conditions are right, the spores germinate and infect the grass plants. But if the fall remains warm or dry or both, the sclerotia may not germinate to form basidia. Instead, they will germinate directly as mycelium under the snow and cause infection that way. In the latter case, management by eradication would be possible, for reinfestation would be a slow process. Nonetheless, *Typhula* blight is a serious annual problem in northern areas.

MANAGEMENT

Cultural Practices

It is important that the turf not be lush going into the winter; if it is, *Typhula* blight will be much worse. It is very difficult to give a date for the last application of nitrogen to actively growing turf because dates for the first lasting snowfall vary so much. Depending on the area, the last nitrogen application should be sometime between mid-August and mid-September. This is not to be confused with dormant nitrogen feeding; that is, applying nitrogen after the top growth has stopped. Dormant feedings promote early greening-up in the spring and quick recovery of turf damaged by snow mold. Whenever possible, effective fungicides should be applied to turf receiving dormant nitrogen feedings.

Genetic

Beard[3] reported that creeping bentgrass cultivars exhibit varying degrees of susceptibility to *Typhula* blight, but from a practical point of view they all require fungicide treatment.

Vargas et al.[156] reported on the resistance of Kentucky bluegrass cultivars to *Typhula itoana*. Adorno, Monopoly, P-114, Bonnie Blue, and Galaxie were the most resistant, while Fylking, Pennstar, Campus, South Dakota Certified, and Prato were the most susceptible. Because Kentucky bluegrass is not as severely damaged as annual bluegrass and the bentgrasses, it may be possible to alleviate the snow mold problem by using more resistant cultivars in areas where the infestation is severe. Vaartnous and Elliott[151] found Park to be the most resistant to *T. ishikariensis*, and Merion and Nugget the most susceptible. The data of Vargas et al.[156] support their findings, although Park had a heavy infestation of *Microdochium* patch. This information may be useful in areas where *Typhula* blight is the most serious disease and melting-out, necrotic ring spot, summer patch and stripe smut are not a problem. (See the section on Kentucky bluegrass varieties in Chapter 9 and the section on interpreting data in Chapter 1.)

The red fescues are, in general, more resistant to snow mold than are the Kentucky bluegrasses and the bentgrasses. According to Vaartnous and Elliott,[151] Boreal and Reptans are the most resistant to *T. ishikariensis*. Use of the fine-leaf fescues is limited by their susceptibility to *Helminthosporium* diseases.The perennial ryegrasses appear to be the most susceptible of all the turfgrasses to *Typhula* blight, followed closely by the creeping bentgrasses.

Biological

Burpee et al. have been able to biologically manage this disease with a nonpathogenic species of *Typhula* called *Typhula phacorrhiza*.[15] They were able to show if nonpathogenic *T. phacorrhiza* could colonize turf first, then infection by the pathogenic *T. incarnata* and *T. ishikariensis* could be prevented.

Chemical

Chemical management of snow molds is necessary on putting greens, whether they be annual bluegrass or creeping bentgrass. Often money is spent on fungicides to manage summer diseases on creeping bentgrass fairways, but snow mold management is forgotten. *Typhula* blight can then infect and kill creeping bentgrass fairways unless fungicides are used, and annual bluegrass will begin to fill in the bare places during the spring. I am not opposed to annual bluegrass fairways, but those who are should realize that a good preventive fungicide program for snow mold management is necessary to prevent creeping bentgrass fairways from turning into annual bluegrass fairways. I disagree with the idea that it is possible to maintain creeping bentgrass fairways without a fungicide program to manage snow molds. Fungicides are needed even in southern areas where people say, "Oh, I only get about 20% snow mold a year, so it's hardly worth spraying." But 20% a year times 5 years is 100%. While all of the creeping bentgrass will not be destroyed in 5 years, most of it will, and annual bluegrass will replace it.

The mercury fungicides have, for many years, been the standard for snow mold management. They will no longer be available for sale after 1994. Although fungicides containing PCNB and chloroneb are also being used now, they are not without problems. PCNB is phytotoxic to some bentgrasses, among them Cohanse (C-7). Toronto (C-15) and Penncross, however, appear to tolerate PCNB. Chloroneb is not effective against *Microdochium* patch, and in areas where both *Typhula* blight and *Microdochium* patch occur or where *Microdochium* patch is the major problem, other fungicides must be used in combination with chloroneb. Furthermore, chloroneb appears to be species-specific, for though it controls *Typhula* blight caused by *T. incarnata*, it is not very effective against *T. ishikariensis*. Vargas and Turgeon[160] showed that chloroneb is a systemic fungicide, which may account for some of the erratic results reported by various researchers, although it is effective as a contact fungicide as well. Stienstra[139] has found that in many areas (and particularly in Minnesota), combinations of two or three fungicides are necessary for snow mold management. This is because many different species of fungi are involved in the snow mold complex.

The DMI fungicides, like fenarimol, triadimefon, and propiconazole, are labeled for snow mold, and in areas like the eastern U.S. where only *T. incarnata* occurs, they are effective. However, they are not effective against *T. ishikariensis*. In areas where both species occur, the DMI fungicides will not effectively manage *Typhula* blight. Combination of iprodione and chlorothalonil has been shown to effectively manage both species of *Typhula* and *M. nivale* in areas where snow covers exist for the winter period.

DISEASE: *Microdochium* patch (pink snow mold)

Pathogen: *Microdochium nivale* (Fr.) Samuels & I. C. Hallett (syn. *Fusarium nivale* Ces. ex Berl. & Voglino)

Hosts: Annual bluegrass, Kentucky bluegrass, rough bluegrass, colonial bentgrass, velvet bentgrass, creeping bentgrass, perennial ryegrass, fine-leaf fescues, tall fescue

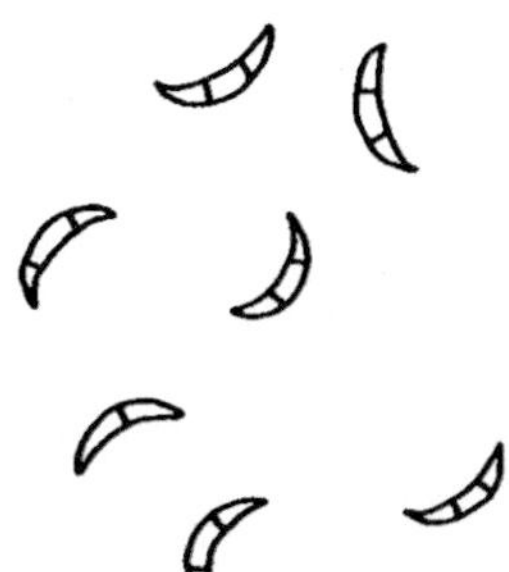

Microdochium nivale

In the Pacific Northwest, the British Isles, and northern Europe, *Microdochium* patch is the most important turfgrass disease. This is mainly because all of these areas have extended periods of cool, wet weather without snow cover. *Microdochium* patch is also a problem in any area with a cool, wet spring or fall or snow cover in the winter months. If annual bluegrass, creeping bentgrass, velvet bentgrass, colonial bentgrass, and perennial ryegrass are infected with *Microdochium* patch, the grass plants may be destroyed. Kentucky bluegrass and red fescue, while susceptible to the disease, are not as seriously affected and the grass plants usually do not die.

Symptoms

Without snow cover, *Microdochium* patch occurs as reddish-brown spots in the turf. The spots range in diameter from less than 1 inch (2.5 cm) to about 8 inches (20 cm), although larger ones are sometimes found. When there is snow cover, the circular spots are usually from 2–3 inches (5 to 7.6 cm) to 1–2 feet (.3 to .6 m) in diameter, and are tan to whitish-gray or reddish-brown (see Figure 4.7). Shortly after the snow has melted, the pink mycelium of the fungus may he seen at the advancing edge of the spot; hence the common name, "pink snow mold."

Figure 4.7. The appearance of *Microdochium* patch on a green after the snow has melted.

Occurrence

Microdochium nivale can survive as mycelium and spores in the thatch, and will actively grow on the grass residue. Infection takes place when the temperature is below 60°F (15.5°C), even though in culture, the fungus grows best at temperatures of 60° (15.5°C) and 70°F (21.1°C).[13] This is probably because when the temperature is below 60°F (15.5°C), the plant grows more slowly and is, therefore, more vulnerable to attack by the fungus. Also, the fungus has less competition from other antagonistic microorganisms at this temperature. *M. nivale* and other snow mold fungi are examples of pathogens that have found ecological niches where they can survive with less competition from other saprophytic and pathogenic microorganisms.

MANAGEMENT

Cultural

Fertilizing with nitrogen causes lush growth going into the snow mold season and will make the turf more susceptible to *Microdochium* patch. It can make management with fungicides more difficult. It is hard to recommend a specific date for the last application of nitrogen, since temperatures and seasons vary so much. It should be early enough to give the turfgrass a chance to harden off before the onslaught of snow or frost or, in areas of mild winters, before the grass becomes dormant. Also, if the turf is lush and there is an open winter (i.e., one without snow cover), large areas of grass can be lost to desiccation. If desiccation covers are used to prevent desiccation, extra fungicide treatments may be needed, as the disease is usually more severe under covers.

Genetic

There is no creeping bentgrass that is truly resistant to *Microdochium* patch; all require preventive chemical treatment. Annual bluegrass also requires chemical treatment. Kentucky bluegrass and annual bluegrass appear to be more susceptible to *Microdochium* patch than is creeping bentgrass. In fact, *Microdochium* patch has been observed to select the annual bluegrass patches on greens that are predominantly creeping bentgrass.

Chemical

There are many contact fungicides that can be used to manage *Microdochium* patch where there is no lasting snow cover. If there is permanent snow cover for three months or longer, either the mercury or the systemic fungicides must be used. Although the benzimidazole systemic fungicides do manage *Microdochium* patch, they also predispose the grass to invasion by *Helminthosporium*, and must be used in combination with contact fungicides, like mancozeb or chlorothalonil. Also, strains of *M. nivale* resistant to the benzimidazole fungicides have been reported in the Pacific Northwest.[17]

Other Snow Molds

Sclerotinia borealis, the low-temperature basidiomycetes (LTB), and a white isolate of *Rhizoctonia solani* are also known to cause snow mold diseases. In Canada, the distribution of the snow mold fungi is *Sclerotinia borealis* in the colder regions, *Typhula* species in the intermediate temperature zone, and *Microdochium nivale* in the warmer regions.[94,95]

The symptoms of LTB are similar to those of gray snow mold except that there are no sclerotia, and many isolates produce hydrogen cyanide. LTB can attack Kentucky bluegrass, red fescue, and creeping bentgrass. PCNB, chloroneb, and chlorothalonil all give effective management.[136]

Snow mold, caused by *Sclerotinia borealis*, has been reported in Canada, Alaska, Sweden, Norway, Finland, and the former USSR. It has been found on Kentucky bluegrass, red fescue, perennial ryegrass, tall fescue,[137] and bentgrass.[136] The disease is similar to other gray snow molds, with grayish mycelium present near the advancing snow margins. Dull black sclerotia can be found on the surface of infected leaves. The disease is most severe in the northern part of Canada.[95] PCNB is an effective fungicide for the management of *Sclerotinia borealis*.

THE SMUTS

DISEASES: Flag smut and stripe smut

Pathogens: Flag smut – *Urocystis agropyri* (Preuss) Schröt
Stripe smut – *Ustilago striiformis* (Westend.) Niessl

Hosts: Kentucky bluegrass, creeping bentgrass, colonial bentgrass

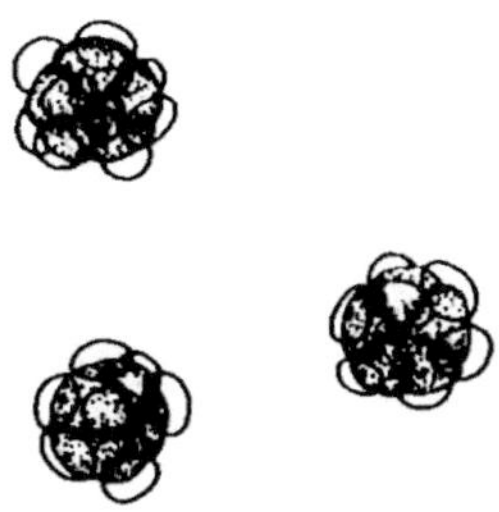

Urocystis agropyri teliospores

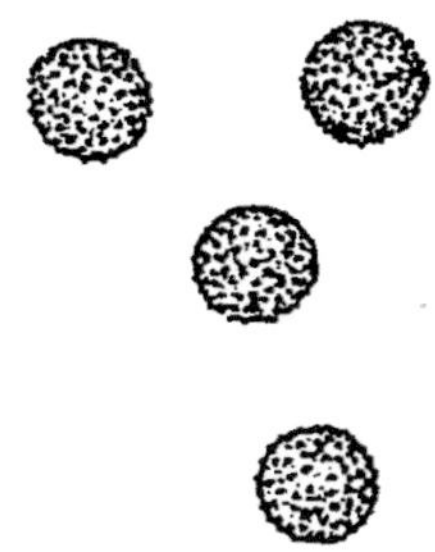

Ustilago striiformis teliospores

The smuts, along with necrotic ring spot and summer patch, are the most destructive diseases of melting-out resistant Kentucky bluegrass cultivars. Smuts are systemic and perennial, which means that once a plant becomes infected it will remain so for life. Furthermore, all plants arising from an infected mother plant via rhizomes will be infected as well. A smut-infected plant is always in a weakened condition and any additional stress may kill it. The most common form of stress is drought, and it is not unusual to see entire lawns that are heavily infected with stripe or flag smut die when people go on vacation in summer and neglect to have someone water the grass. The smuts can be distinguished easily under a microscope. The teliospores of stripe smut occur singly, while those of flag smut occur together in groups of three to five, known as "spore balls," that are surrounded by sterile cells.

The following discussion is mainly about stripe smut on Kentucky bluegrass because little work has been done on flag smut, although flag smut may be the more important disease of the two. Plants infected with flag smut are killed more readily than are plants infected with stripe smut. Consequently, a year after infection has occurred it is difficult to find plants that have flag smut, but plants affected by the less severe stripe smut remain (C. F. Hodges, personal communication).

Stripe Smut on Bentgrass

Although most of the attention has been given to stripe smut on Kentucky bluegrass, stripe smut can be a serious problem on creeping bentgrass as well. It is often difficult to see smut on bentgrass cut at putting-green height, because there is so little leaf surface. Stripe smut has been reported on the cultivars Evansville, Pennlu, Seaside, Toronto (C-15), Arlington (C-1), Congressional (C-19), Old Orchard (C-52), and Penncross.[58] I have observed stripe smut to be a problem mainly on Toronto creeping bentgrass in the Midwest. If it occurs on Penncross, that is the exception rather than the rule. The disease usually starts thinning the turf in high spots or heavily trafficked areas (e.g., where the cup is cut). Recovery of the turf around the cup following a day with heavy play will be slow. Annual bluegrass rather quickly invades such weakened areas on Toronto creeping bentgrass greens. Stripe smut is a slow-moving disease, which is difficult to detect in Toronto creeping bentgrass greens because of the height at which they are mowed, and except for the thinning, slow growth, and invasion of annual bluegrass, it may go unnoticed.

Symptoms

From a distance, infected turf looks clumpy because the stripe smut fungus causes the infected bluegrass plants to grow more upright than the healthy plants.[70] Stripe smut tends to kill individual plants, leaving bare spots that contribute to the clumpy appearance. Smut-infected plants do not tiller as profusely as healthy plants, and so weed grass and broadleaf weeds, rather than Kentucky bluegrass, fill the bare spots.

Light yellow blades of grass are the first symptoms of stripe smut. As the disease advances, the leaf blades begin to curl and show parallel black stripes running the length of the blade (see Figure 4.8). If you touch these stripes, a black, sootlike dust—the spores of the stripe or flag smut fungus —will rub off. Older infected blades will be twisted, curled, and shredded from the tips down.

Occurrence

The stripe smut fungus overwinters as mycelia in the crown and other vegetative parts of infected plants. It can also survive as teliospores in the soil for many seasons. Infection of seedlings may occur through the coleoptiles or through axillary crown buds or rhizome and stolon nodes.[71]

Figure 4.8. Close-up of Kentucky bluegrass leaf blades infected with stripe smut.

Symptoms of stripe smut are most commonly seen in the spring and fall during periods of wet, cool weather, when the daytime temperature is below 70°F (21.1°C). They gradually disappear as the weather becomes warmer. Although the symptoms are most evident during cool weather, very little turf is lost then. Most turf loss occurs during hot, dry summer weather (when the grass is under heat and drought stress), or in open winters that subject plants to desiccation and cold-temperature stress.

Stripe smut is a systemic disease that is perennial in the grass plant. *Systemic* means that the fungus is internal and can spread throughout the plant's vascular system. (The striping on the grass blades is due to the fungus growing only in the veins.) *Perennial* means that once a plant is infected it will remain so for life, although symptoms may not always be visible. Under certain environmental conditions, foliar symptoms may be absent even though the fungus is still present in the crown of the infected plant. Protected in the crown, the fungus can survive adverse conditions for months and then become active when conditions are favorable.

The stripe smut fungus cannot attack through the foliage; it can attack only through the lateral buds on the rhizomes, but leaves, tillers, and rhizomes arising from infected crowns are infected also. This limited ability to infect probably explains why the disease is usually serious only on turf that is at least 3 years old. It is also the reason that stripe smut becomes a problem in sod fields that have been in cultivation for many years, allowing spores to accumulate in the soil. The spores are thick walled and can remain alive in the soil for many years.

Stripe smut produces specific races that attack specific cultivars of Kentucky bluegrass. Gaskin[59] was the first to demonstrate the existence of these races. All races of stripe smut belong to the species *Ustilago striiformis*, but each has unique characteristics resulting from its own particular genetic makeup. Each race of stripe smut will attack only specific Kentucky bluegrass cultivars.

The spores of *U. striiformis* are often referred to as chlamydospores because they are thick-walled and can survive in the soil for many years, but they are more correctly called teliospores. The sexual spores undergo typical genetic recombination before forming the thin-walled spores that actually cause new infection. As a result, the potential for new races of *U. striiformis* exist with each generation. Therefore, the fact that a cultivar is resistant today doesn't mean it will be resistant forever. A "resistant" cultivar is simply one that is resistant to the stripe smut races prevalent today.

MANAGEMENT

Cultural Practices

Keep nitrogen applications to a minimum during the summer months—no more than one-half pound (.25 kg) of nitrogen per 1000 square feet (100 square meters) per month. Turf infected with stripe smut should not be allowed to dry out. A healthy Kentucky bluegrass turf will become dormant when dry and will recover when it is irrigated, but a lawn with stripe smut will die no matter how much you water it.

Genetic

There are several Kentucky bluegrass cultivars that are resistant to stripe smut, but their resistance is likely to be temporary. The stripe smut fungus has the potential to produce new races, and once a Kentucky bluegrass becomes widely grown, a race of *U. striiformis* which can attack it will probably develop. Cultivars that are currently resistant are certainly preferable to cultivars that are already susceptible, but you should be on the lookout for problems in the future. What is needed is identification of Kentucky bluegrass cultivars that have horizontal resistance to stripe smut (see Chapter 8). In the meantime, it is a good idea to blend three or four of the better Kentucky bluegrasses. Then, if a new race of stripe smut that is specific to one of the cultivars in the blend develops, the other cultivars will remain healthy, and the turf will not be ruined.

The best way to prevent stripe smut from destroying a home lawn or general turf area is to plant blends of Kentucky bluegrass cultivars in the home lawn or general turf areas. Golf course turfs should be seeded to a blend of creeping bentgrass to prevent stripe smut from destroying them.

Chemical

Stripe smut is an internal disease and can be managed chemically only by the use of systemic fungicides. The DMI fungicides, fenarimol, triadimefon and propiconazole, are the most effective in managing stripe smut. Making monthly applications during the growing season will manage stripe smut. While this may be expensive, so is renovating a green. For the home owner where monthly fungicide applications may be impractical, renovation and planting a blend of improved Kentucky bluegrass may be the answer.

THE RUSTS

Disease: Rusts

Pathogen: *Puccinia* spp.

Hosts: *Puccinia graminis* Pers.: Pers subsp. *graminicola* Z. Urban -Stem rust, Kentucky bluegrass
Puccinia striiformis West. -Stripe rust, Kentucky bluegrass
Puccinia coronata Corda f.sp. *agropyri* Erikss. -Crown rust, perennial ryegrass, tall fescue
Puccinia cynodontis Lacroix -Bermudagrass rust
Puccinia zoysia Diet. -Zoysiagrass rust
Puccinia brachypodii Otth. var. *poae-nemoralis* (Otth.) Cummins and H.C. Greene -Leaf rust, Kentucky bluegrass

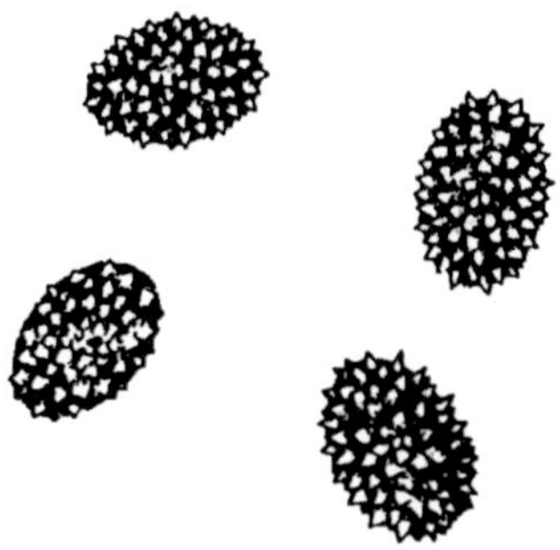

Puccinia graminis urediospores

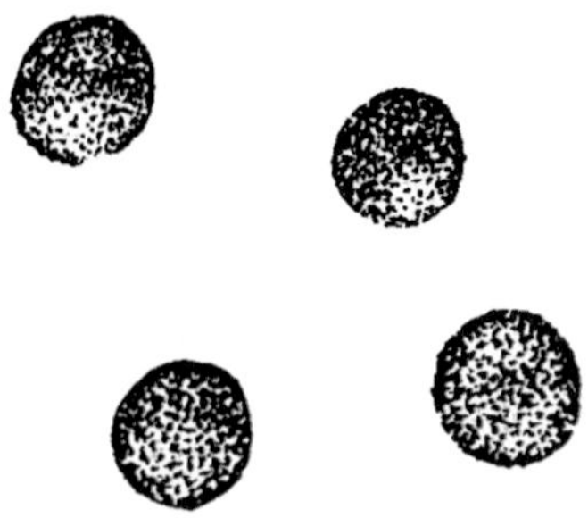

Puccinia coronata urediospores

Puccinia graminis teliospores

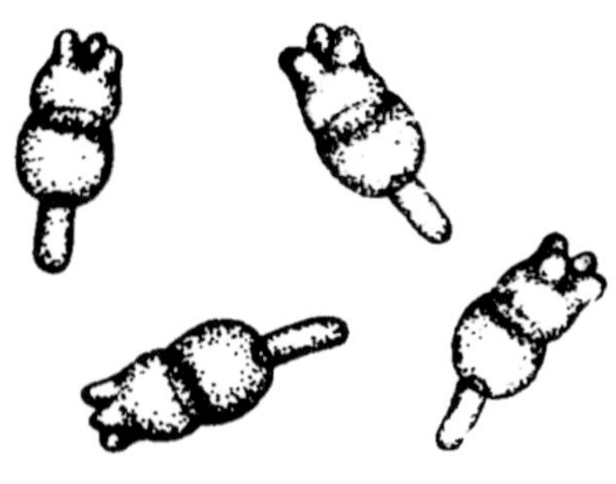

Puccinia coronata teliospores

The rusts are, in general, minor diseases on mowed turf; the one exception is stripe rust. On adequately fertilized turf that is mowed once a week, rust is not a problem. It can become a problem in coastal areas where slow growth occurs because of cool weather or if cool weather comes early in the fall, rust can become serious enough to warrant the application of fungicides. Stem rust on Kentucky bluegrass and crown rust on perennial ryegrass are serious problems in seed fields in the Pacific Northwest; such fields do require fungicide treatment.

Symptoms

The rust diseases are characterized by the yellow to dark brown urediospores that protrude through the ruptured epidermis of the grass plant. At a distance rust-infected turf looks yellow, but close examination will reveal rust-colored urediospores protruding through the epidermis of the grass plant.

Occurrence

Stem Rust

Rusts have different spore stages which occur on different hosts. *Puccinia graminis* has five spore stages: the uredial and telial, which occur on the grass; the basidiospores, which are produced on the teliospores and then infect the barberry; the pycnial stage, which occurs on the barberry and during which mating takes place; and the aecial stage, when the aecial spores are produced and then carried by the wind back to the grass host. All this is interesting, but has very little practical value in the epidemiology of the disease.

In the warmer portions of the Kentucky bluegrass growing region, the fungus can overwinter as mycelium or active urediospores in infected plants. Britton[12] found stem rust surviving as far north as Lafayette, Indiana. In the spring, urediospores are produced on overwintering plants and are carried northward on the prevailing southerly winds of summertime. They land on susceptible Kentucky bluegrass plants, germinate, and infect the plants. It takes 10 to 14 days from the time of infection until the spores erupt through the epidermis. Each infecting urediospore can produce hundreds of new urediospores, and it doesn't take long for an epidemic to develop.

Leaf Rust

Leaf rust can overwinter in cool regions on Kentucky bluegrass plants. It is not as widespread or as important as stem rust.

Stripe Rust

Lawns severely infected with stripe rust look orange or brown. The pustules in the leaves are yellow and run lengthwise up the sheath and leaf blade between the vein. On the West Coast of the U.S., stripe rust is active during the winter months[47] and is most severe during cool moist periods in the early spring and late fall. The disease can cause serious loss of turf in the western United States.

Zoysiagrass Rust

The rust on zoysiagrass was first described by Kreitlow et al. in 1956.[85] It can be a serious problem because zoysiagrass grows slowly and consequently is mowed less frequently than are other turfgrass species. Zoysiagrass lawns infected with *Puccinia zoysia* have an orange discoloration and may be severely thinned. The disease is most serious during periods of cool moist weather and in shaded areas.[57,84,85] Emerald and Meyer are the most susceptible cultivars.[57,84] Severe rust infection on zoysiagrass usually requires fungicide treatment. Maneb and sulfur both control the disease when applied at 5-day intervals.

Crown Rust

Crown rust, caused by *Puccinia coronata*, is a common disease on perennial ryegrass and tall fescue. The pustule is yellow and the disease will turn the turf the same color. The disease gets its name from the crownlike appearance of the teliospores. Pennfine and Manhattan are very susceptible and crown rust is becoming a limiting factor in their use, especially in the southern range and coastal areas of the United States. One of the main uses of perennial ryegrass is for athletic turfs, notably football fields. The rusts are worst when the grass is growing the slowest, which is exactly the time that football is played. Where rust is a problem, both Pennfine and Manhattan should be avoided.

Bermudagrass Rust

Puccinia cynodontis can cause a serious rust problem on some of the improved bermudagrass cultivars, although it seldom is severe enough to destroy a bermudagrass turf. In nurseries where bermudagrass cultivars are grown to be harvested and sold as sprigs for propagation, *P. cynodontis* can do a lot of damage. In nurseries the bermudagrass is not mowed, and consequently the rust has a chance to go through many infection cycles. The cultivar Sunturf is one of the most susceptible bermudagrasses, but the common types are resistant.[162] Infected grass blades have pustules that contain cinnamon-brown urediospores and black teliospores, but the general appearance of rust-infected bermudagrass turf is yellow.

MANAGEMENT

Cultural

In many cases proper use of nitrogen can eliminate rust as a problem. If the turf can be mowed at least every 7 days, rust should not be severe. Zoysiagrass, which is slow-growing, is an exception. Fungicide treatments may be necessary on other turfgrass species as well as in areas where the turf grows slowly because the spring and fall are prolonged, cool, and wet. Bear in mind that applying excessive nitrogen in the fall to control rust could lead to more serious problems with snow molds or desiccation. During this time of year, fungicides are a better treatment for minor rust problems than is nitrogen, since high levels of nitrogen encourage other diseases that are more damaging than rust.

Genetic

The rusts are most severe on improved cultivars and almost nonexistent on the common types of grasses. More explicitly, cultivars like Merion Kentucky bluegrass, Emerald and Meyer zoysiagrasses, Sunturf bermudagrass, and Manhattan perennial ryegrass, which each consist of a single genotype, are very susceptible, while the cultivars known as common types (e.g., common bermudagrass, common perennial ryegrass, common Kentucky bluegrass) are natural blends of different genotypes of the same species. Rusts produce specific races that attack specific grass cultivars[162] (see section on physiologic races in Chapter 1). If a turfgrass community is composed of grass of a single genotype, a race of rust specific to that genotype can destroy the entire turf. On the other hand, if a turf is a blend of grass plants of differing genotypes, a specific race of rust that attacks one genotype probably will not harm the others. Instead of the whole turf being destroyed, the worst that might happen would be loss of one of the grasses making up the blend.

Blending some of the improved cultivars would be a way to avoid severe outbreaks of rust. But why not simply plant a common type? Because the most important disease must be managed first and by the most efficient means. In the case of Kentucky bluegrass that means managing melting-out, summer patch, necrotic ring spot, and stripe smut. Planting common, Newport, or Park Kentucky bluegrass instead of a cultivar resistant to all three of these major diseases, just to avoid rust, would be like catching smallpox just to avoid cowpox! It is far easier to manage mild diseases like rust than devastating ones like melting-out, summer patch, necrotic ring spot, and stripe smut. Planting blends of some of the newer Kentucky bluegrass cultivars that are resistant to all three diseases should reduce the rust problem. So far there are no conclusive data on the resistance or susceptibility of the new improved Kentucky bluegrass cultivars to rust. Knowing the rust fungus, with its numerous races, I would guess that all cultivars will eventually become susceptible. For a further discussion of blending, see Chapter 8.

Chemical

As explained before, chemical management of rusts may be necessary on slow-growing turf. Chemical management is also needed on grasses grown for seed production, because they are not mowed, and on cultivars propagated vegetatively (e.g., some cultivars of bermudagrass and zoysiagrass), which are not mowed either.

CHAPTER 5

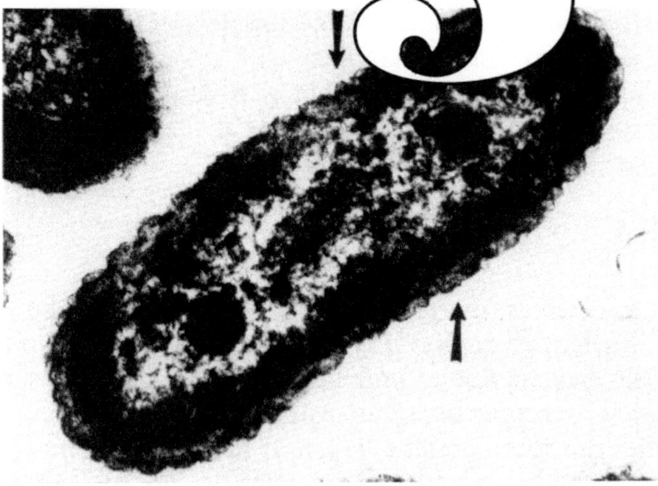

Bacterial, Mycoplasmal, Nematode Diseases of Turfgrass, and the Black Layer

BACTERIAL DISEASE

DISEASE: Bacterial Wilt

Pathogen: *Xanthomonas campestris*

Hosts: Creeping bentgrass [cultivars-Toronto (C-15), Nimisilla, Cohansey (C-7)], bermudagrass, Kentucky bluegrass, annual bluegrass, and perennial ryegrass

This disease is most remembered for the devastation it caused to Toronto (C-15) creeping bentgrass greens in the Midwest during the early 1980s. In fact, the disease is still referred to as C-15 decline. However, the proper name is bacterial wilt, as described by Roberts et al.[122] The disease had been observed for many years on Toronto creeping bentgrass greens, but always at low levels. The disease was initially diagnosed incorrectly as red leaf spot[102] and later as leaf blight and crown rot.[89]

The severe outbreak of bacterial wilt has been associated with the switch from standard soil topdressing of greens to the use of the light frequent sand topdressings programs which began in the early 1980s. Whether it was the wounding of the plants by the sand that allowed entry of the bacterium into the plant, or the low level of beneficial microorganisms in a sand medium to compete with the pathogen, the answer is not known. What is known is that golf course after golf course lost their Toronto creeping bentgrass greens as they switched to a sand topdressing program.

Symptoms

The disease first appears as tiny red- to copper-colored spots about the size of a dime. It often resembles tiny dollar spots, except for the color. The size of the spots is due to individual plants dying, and the color is characteristic of creeping bentgrass wilting. As more and more individual plants die, the spots become larger. If ideal conditions for disease development persist, an entire green may be lost in a short period of time (Figure 5.1).

Figure 5.1. Bacterial wilt severely thinning a Toronto creeping bentgrass green. Notice patch of annual bluegrass beginning to invade.

Occurrence

Bacterial wilt begins to develop during extended periods of rainfall. The disease can be devastating if the prolonged period of rain is followed by a period of bright sunny days and warm weather. Symptoms appear to be related to the plugging of the xylem tissue of the grass plant by the bacterium (Figure 5.2). As stated before, the disease is most devastating where sand topdressing programs are followed. In the United States the disease has been a problem on the vegetatively propagated creeping bentgrass cultivars, i.e., Toronto, Nimisilla, Cohansey, and on the vegetatively propagated bermudagrass cultivar Tifgreen. The reason it became so widespread was probably due to the bacterium becoming established in the nurseries where these cultivars were propagated. The creeping bentgrass cultivars infected with *X. campestris* were then transferred from these nurseries to the golf courses as stolons, sprigs, or sod.

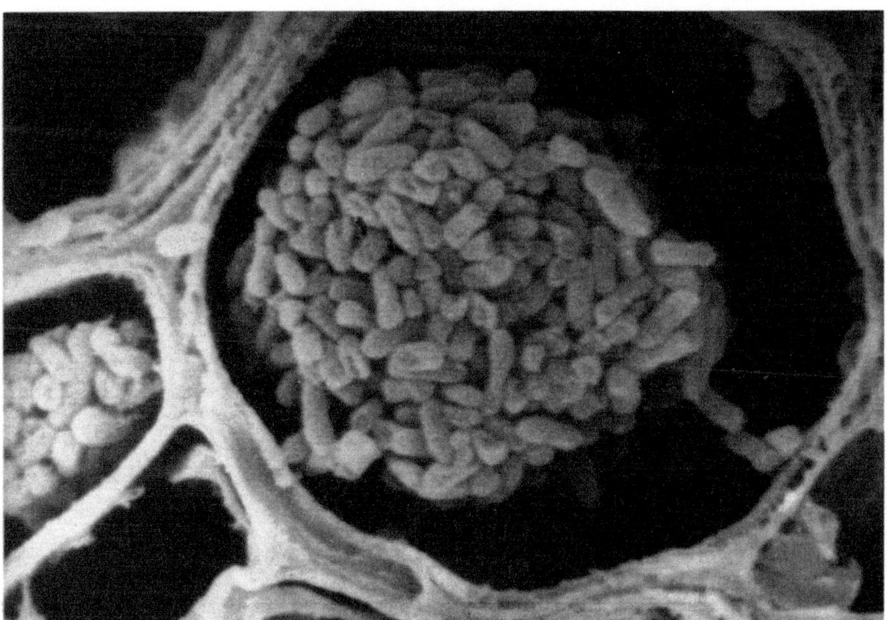

Figure 5.2. Electron photomicrograph of *Xanthomonas campestris* in xylem vessel of a creeping bentgrass root.

MANAGEMENT

Cultural Practices

There is very little that can be done culturally to manage this disease, other than to avoid going to a sand topdressing program where this disease is present.

Chemical

Roberts et al.[123] indicated that the antibiotic, oxytetracycline, at heavy rates was shown to manage bacterial wilt for a 4- to 6-week period (Figure 5.3). This is a very costly and time-consuming process because of the high rates of oxytetracycline and the large volumes of water that must be used. This is more of a temporary solution to get through a golfing season. The only real solution to the problem lies in regrassing the greens to resistant cultivars.

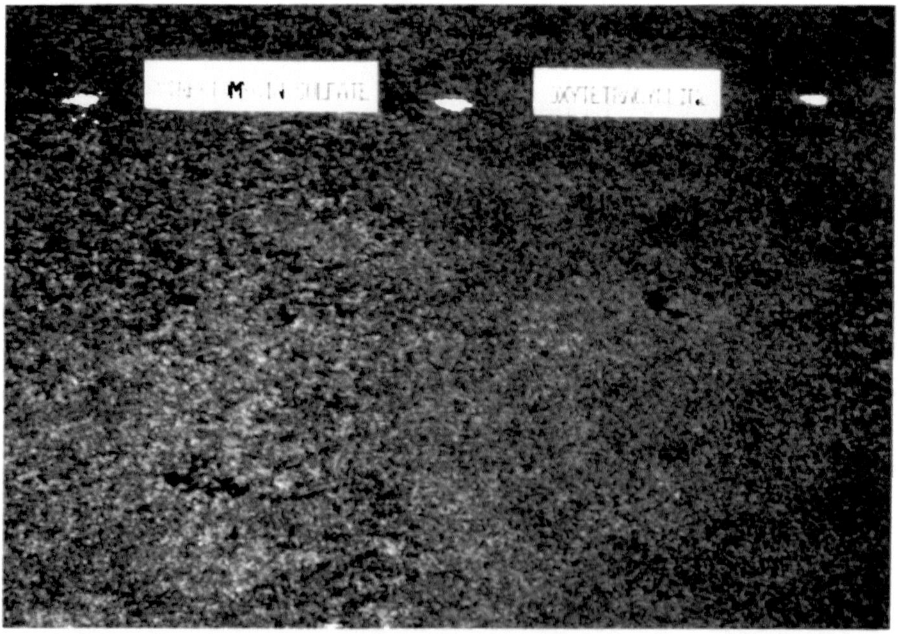

Figure 5.3. Oxytetracycline controlling bacterial wilt in a Toronto creeping bentgrass green.

Genetic

All the seeded cultivars of creeping bentgrass have so far been resistant to bacterial wilt. If they do become susceptible, it will probably be isolated to a few golf courses unless bacterial wilt becomes a serious problem in the seed fields. Neither is highly likely, because of the genetic diversity that exists in the newly seeded creeping bentgrass cultivars.

VIRAL AND MYCOPLASMAL DISEASES

There are 24 viral diseases and 1 mycoplasma-like disease known to affect turfgrass.[131] Of these, St. Augustine decline (SAD) is the only one of economic importance. This may be the true situation; or it could be that there are viral problems in many of the important turfgrass species but the plants are symptomless carriers; or viruses could be expressing themselves as part of a complex of problems that might include edaphic factors, fungi, nematodes, and various environmental stresses. Control of one or more aspects of such a complex might lessen the effect of a virus and cause it to go unnoticed. The greatest potential for harm from viruses is probably in seed-production fields. Many viral diseases of other grasses (wheat, barley, oats, etc.) tend not to destroy the host but merely to reduce yields.

The St. Augustine decline virus is the only one discussed in detail, since it is the only viral disease of turfgrass that is economically important. For information on the other viral diseases and the mycoplasma-like disease, consult the book by Smiley et al.[131]

The pathogen responsible for SAD is the Panicum mosaic virus, or St. Augustine decline virus.[98] The host, St. Augustinegrass, is one of the major turfgrass species in the South. In Texas and Florida, SAD damages or destroys more St. Augustinegrass than any other disease. An estimated 3 million dollars in turf replacement value was lost in Corpus Christi, Texas, in 1966.

The first symptom of SAD is mild chlorotic mottling of the leaf blades (Figure 5.4). The mottling becomes worse in the second year, and in the third year, infected grass plants may eventually die, leaving dead areas that are filled in by weedy grass and broadleaf weeds. The disease is most severe when St. Augustinegrass is grown under conditions of low nitrogen or drought, or where nematodes or insects are a problem.[98]

SAD is mechanically transmitted. Since all the St. Augustinegrass

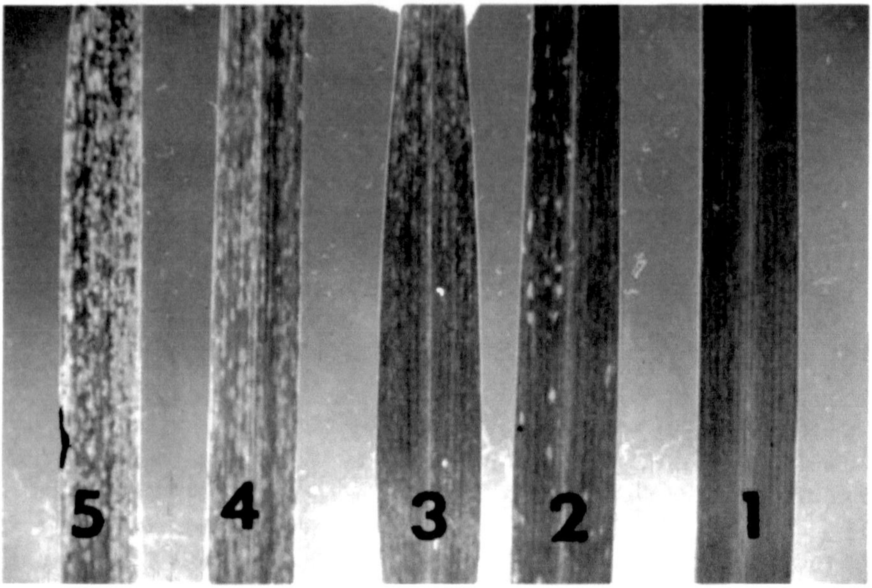

Figure 5.4. **Foliar symptoms of St. Augustine decline (SAD). 1 = no disease, 5 = severe disease. (Photograph courtesy of Dr. Don Blasingame.)**

turfs are sodded, sprigged, or stolonized (sprigged being the most common method of grass establishment), once the disease becomes established in a St. Augustine nursery it can be spread throughout the nursery by mowing, and over a large area when the sprigs are sold. Likewise, once SAD gets into a neighborhood where the mowing is done by contractors, it can spread quickly from lawn to lawn.

MANAGEMENT

Cultural

Good turf maintenance will slow the progress of the disease, but the decline and eventual loss of St. Augustinegrass turf is inevitable. Nitrogen applications can temporarily mask the presence of SAD, but in time the symptoms will reappear.

Genetic

Texas A & M University, in cooperation with the University of Florida, has developed a cultivar of St. Augustinegrass called Floratam, which is resistant to SAD. This, of course, sounds like the obvious solution to the SAD problem, both for new lawns and for reestablishing St. Augustinegrass in turfs destroyed by SAD. In addition to its resistance to SAD, Floratam is resistant to the chinch bug, which also is devastating to St. Augustinegrass. Unfortunately, Floratam lacks low temperature hardiness which makes it an undesirable turfgrass species, especially in the cooler parts of the St. Augustinegrass growing area. Also viruses, like the rust and smut fungi are highly mutable, and have the potential to overcome host resistance.

Chemical

There are no chemicals that will manage SAD.

NEMATODES

Nematodes are small nonsegmented round worms, usually eel-shaped, which live in soil and water or on hosts (plants or animals). Plant parasitic nematodes are colorless and range from 1/10 to 1/75 of an inch (.25 to .03 cm) in length (see Figure 5.5). All plant parasite nematodes are obligate parasites. They feed by means of a stylet, a hollow, spearlike structure located in the anterior (head) end (Figure 5.6). With

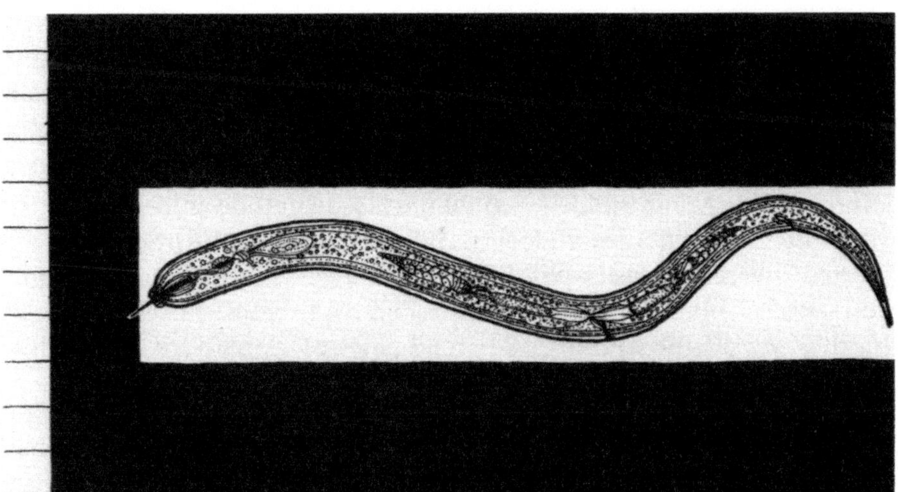

Figure 5.5. Ectoparasitic nematode and No. 50 thread (black background), both at a magnification of 270X. The common button thread whose diameter is approximately 12 times that of the worm, is visible to the naked eye; the nematode is microscopic and invisible.

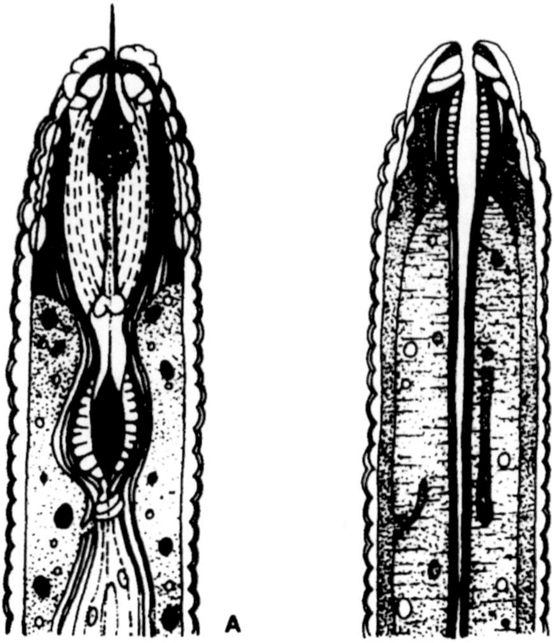

Figure 5.6. *A*, **plant-parasitic nematode with stylet;** *B*, **free-living nematode without stylet.**

its stylet, the nematode punctures living cells to obtain food. The stylet is like a drinking straw through which the nematode draws a cell's contents into its own body. They damage the roots of the plant, thus interfering with the plant's ability to take up water and nutrients. Also, before withdrawing the cell's contents the nematode injects digestive secretions that start decomposing the cell's protoplasm. Some of the injected substance causes physiological changes that make grass plants more susceptible to other problems, such as diseases caused by fungi. In other words, the wounds that some types of nematodes make while feeding may allow other parasites or saprophytes to enter.

The two main classifications of nematodes are based on their feeding habits. Ectoparasitic nematodes (Figure 5.7) feed with their bodies outside the plant root, and endoparasitic nematodes (Figure 5.8) enter the plant root and feed from within the root. Ectoparasitic nematodes are more common on turfgrass, which is fortunate because they are easier to manage.

Figure 5.7. Ectoparasitic nematode feeding on a turfgrass root.

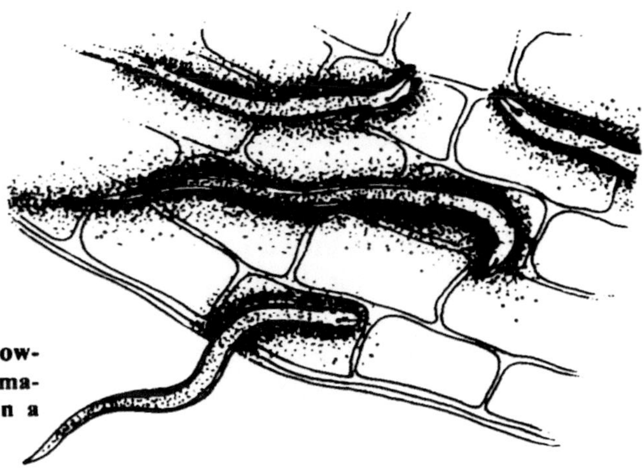

Figure 5.8. Endoparasitic (burrowing or lesion) nematodes feeding within a turfgrass root.

Pathogenicity

Plant parasitic nematodes that live in a weed-free stand of turf must feed on the roots of the turfgrass. The question that has been baffling turfgrass scientists is, "How much damage are they doing?" A single nematode has negligible potential for doing damage, but it doesn't take long for nematode populations to reach a level where they can do great harm. The life cycle (i.e., egg to egg) of plant parasitic nematodes may be as short as 10 days, but 30 to 45 days is normal. Depending on the species, one female can lay up to 2000 eggs. Nematode populations vary throughout the season. If you suspect that nematodes are present, take samples every month throughout the growing season to see how the population fluctuates. This information will help you work out strategies to control the problem with nematicides.

How many nematodes constitute a problem? The answer depends on the species of nematode. As few as 10 sting nematodes (*Belonolaimus longicaudatus*) per 100 cubic centimeters of soil are a real problem, whereas 300 stunt nematodes (*Tylenchorhynchus* spp.) per 100 cubic centimeters are a potential problem.[92] The data to support these numbers come from laboratory and greenhouse experiments carried out in controlled environments with pasteurized soil to eliminate other biological factors such as microorganisms, insects, and even other nematode species. It is understood that the direct application of the results of laboratory and greenhouse experiments to field conditions is not always possible, but these data can help you decide how to approach field situations.

How many nematodes are necessary to cause damage when soil is water-soaked following heavy rains? How many are necessary to weaken a plant or alter its physiology so that weak pathogens like *Rhizoctonia cerealis* or *Pythium graminicola* can become serious? How many nematodes constitute a problem when more than one species is found in association with a plant? Do 300 nematodes per 100 cubic centimeters of soil have the same effect on a grass plant growing at 70°F (21.1°C) as they do on a grass plant growing at 90°F (32.2°C) or more? The point is that the pathogenic potential of a nematode population is influenced by a complex of biotic and abiotic factors. Laboratory experiments elucidate individual components of the complex, but all aspects of the situation must be taken into consideration when you make decisions about a particular field situation.

Sampling

Your sampling technique is important if laboratory estimates of the nematode populations are to be accurate. Collect a pint (.5 liter) of soil by taking random cores from an area suspected of harboring nematodes. Put it in a container, preferably a plastic bag, label it properly, and take it to a test laboratory immediately. If this is not possible, place the sample in a refrigerator (not a freezer) until it can be taken to a laboratory. Many nematodes will die if subjected to high temperatures or drying (for example, being left on a desk or in a truck for a day or two) and the information you receive from the laboratory may not reflect the true population.

Inaccurate and misleading information can also result from sampling too late; that is, sampling an area so badly deteriorated that there are only a few live plants left. Nematodes are obligate parasites and once their food source, the plant roots, are destroyed, the population declines. This could lead to an erroneous conclusion that nematodes aren't the problem. If much turf has already been killed, it is advisable to collect soil samples from adjacent healthy areas. If tests show high nematode populations in the areas where the turf looks healthy, you must keep a close watch for further signs of deterioration.

Locality

It is easier to recognize nematode problems in the warm-season grasses of the South than in the cool-season grasses of the North. This is primarily due to the longer periods of warm weather in the South. Nematodes are a problem on turfgrass when they cause severe thinning and shortening of the root system (see Figure 5.9). A turfgrass plant with such a root system may survive when it is growing at a cool or moderate temperature, but it will not survive long periods of high temperature. Even when adequate water is available, the damaged root system cannot take up water fast enough to replace the water being lost through transpiration. Heald and Perry[66] point out that nematode problems are most serious in the sandy coastal soils of Florida, which are more subject to drought than are heavier soils. This supports the hypothesis that symptoms are caused by the interaction of nematodes, drought, and high temperature.

In the North, symptoms of nematode problems are more subtle. Because of the more moderate climate, turfgrass plants attacked by nematodes can endure severe damage to the root system and still survive, especially if they get adequate water. Because of its damaged

Figure 5.9. Kentucky bluegrass root shortened by stunt nematode. *Tylenchorhynchus dubius* (left), and a healthy nematode-free root (right).

roots, turf with a high nematode population wilts faster than healthy turf, but in the North, irrigation can rectify this. Sometimes the only noticeable difference between a golf course putting green with a high nematode population and one with few nematodes or none at all is how much faster the green with the high nematode population wilts. Another reason that symptoms in the North are different from those in the South is that the turfgrass and nematode species involved in the host-parasite relationship are different.

Biological interactions usually are not simple, and this is true of nematode problems in turf. In northern turfgrasses, nematodes often interact with other plant pathogens. The more moderate the climate, the more these interactions will affect the symptoms. There are many examples of crops being affected by nematode-fungus interactions, but only a few such interactions have been noted on turf. This may be because the leading turf researchers of the 1950s and early 1960s were mainly concerned with comparing turf treated with nematicides to untreated turf and did not pay much attention to environmental stress, fungal interactions, or microsymptoms. Many laboratories have used nematode extraction techniques that do not adequately assess the total nematode population, and frequently nematodes have been identified only at the generic level, even though species identification is necessary for proper analysis.

Specific Nematodes

Nematodes that affect turfgrass are listed in Table 5.1. Their common and scientific names are given, along with the names of grass species or cultivars on which they are known to be a problem. The sting nematode and the spiral nematode deserve special mention because they are such important parasites of turfgrass.

Sting Nematode (*Belonolaimus* spp.)

The sting nematode is a large ectoparasitic nematode that feeds on the tips of roots. It can cause severe damage in relatively low numbers (less than 10 per 100 cubic centimeters of soil) with its long stylet and the vigorous manner in which it feeds. It is the most important turfgrass pest in Florida. It is known to be a problem on St. Augustinegrass, centipedegrass, and zoysiagrass.[66] Christie et al.[21] described the symptoms produced by the sting nematode as "stubby root" because of the knoblike appearance of the root tips on the few remaining rootlets (Figure 5.14). The short root system may be accompanied by stunted top growth and yellowing.

Spiral Nematode (*Helicotylenchus* spp.)

Perry et al.[112] found *Helicotylenchus digonicus* to be responsible for "summer dormancy," a disease associated with unthrifty Kentucky bluegrass in Wisconsin turf. *H. digonicus* greatly reduced the number and size of the Kentucky bluegrass roots during the spring and early summer, and the symptoms of summer dormancy developed with the arrival of high temperatures and drought stress. The reduction of nematode populations with nematicides improved the quality of the turf and prevented summer dormancy.

The characteristic symptoms of summer dormancy are a thin stand, which is difficult to maintain, curtailment of growth of new turfgrass plants from rhizomes or stolons, and pale or chlorotic narrow leaf blades and brown roots. Generally, the turfgrass plants grow well in the spring, then decline and exhibit the disease symptoms in the summer, and resume good growth in the fall. This complex of environmental stress and disease supports the idea that nematodes have an important role in symptom development.

Table 5.1. Pathogenic Turfgrass Nematodes, Symptoms, and Turfgrass Species With Which They Have Been Associated

Common Name	Scientific Name	Associated Turfgrass Species	Symptoms
Endoparasitic Nematodes			
Burrowing nematodes[20,52]	*Radopholus similus*	Bermudagrass St. Augustinegrass	Yellowing and stunting of top growth; stunting of root system; brown sunken lesions, often quite large.
Cyst nematode[39,40] (Figures 5.10 and 5.11)	*Heterodera major*	Tall fescue Fine-leaf fescue Rough bluegrass Annual bluegrass	
	H. punctata	Creeping bentgrass Colonial bentgrass St. Augustinegrass	
	H. leuceilyma	St. Augustinegrass	
Lesion nematode[147,148]	*Pratylenchus brachyurus*	Bermudagrass Kentucky bluegrass	Yellowing of top growth; large brown lesions on root system; severe pruning of root system.
	P. penetrans	Tall fescue Kentucky bluegrass	
	P. zeae	Zoysiagrass	
	Pratylenchus spp.	Bermudagrass Centipedegrass St. Augustinegrass Zoysiagrass	
Root-knot nematode[64,73,74,113,119,120] (Figure 5.12)	*Meloidogyne arenaria*	Bermudagrass	Yellowing and stunting of top growth; swelling of roots; knots or galls present.
	M. hapla	Bermudagrass Kentucky bluegrass St. Augustinegrass	
	M. incognita	Creeping bentgrass Bermudagrass Kentucky bluegrass	
	M. naasi	Creeping bentgrass Colonial bentgrass Tall fescue Perennial ryegrass Annual bluegrass Kentucky bluegrass Rough bluegrass	

Table 5.1. Continued

Common Name	Scientific Name	Associated Turfgrass Species	Symptoms
Endoparasitic Nematodes (continued)			
	M. graminis	Bermudagrass Kentucky bluegrass St. Augustinegrass Zoysiagrass	Top growth stunted and yellow; wilting evident; roots brown and stunted; root enlargement and galls small and often difficult to see.
Grass root gall nematode[138]	*Subanguina radicola*	Annual bluegrass	Large galls evident on roots.
Ectoparasitic Nematodes			
Awl nematode[110]	*Dolichodorus* spp.	Bermudagrass	Stunted roots and yellow foliage.
Dagger nematode[66]	*Xiphinema americanum*	Bermudagrass Zoysiagrass Kentucky bluegrass Centipedegrass St. Augustinegrass	Yellowing and stunting of top growth; stunted root system with reddish-brown to black lesions.
Lance nematode[39,121]	*Hoplolaimus* spp.	Bermudagrass Annual bluegrass Centipedegrass St. Augustinegrass Zoysiagrass	Stunting of top growth; swelling of roots at feeding areas.
Pin nematode[26,27,100]	*Paratylenchus hamatus*	Kentucky bluegrass Tall fescue	Stunting of top growth with shortened internodes; root systems increase in size with few lateral roots; increased tillering; lesions evident on infected roots.
	P. projectus	Tall fescue Fine-leaf fescue Kentucky bluegrass	
Ring nematode[21,56,83]	*Criconemella cylindricum*	Kentucky bluegrass Zoysiagrass Bermudagrass Centipedegrass	Stunting and thinning of turf. Stunted roots with brown lesions.
	Criconemella spp.	St. Augustinegrass Fine-leaf fescue Annual bluegrass Creeping bentgrass	
Sheath nematode[4]	*Hemicycliophora* spp.		Yellowing of top growth.

Table 5.1. Continued

Common Name	Scientific Name	Associated Turfgrass Species	Symptoms
Ectoparasitic Nematodes (continued)			
Spiral nematode[39,40]	*Helicotylenchus digonicus*	Kentucky bluegrass Bermudagrass	Yellowing and thinning of turf. Stunted discolored root system.
	H. erythrinae[30,92,100]	Creeping bentgrass Kentucky bluegrass St. Augustinegrass Zoysiagrass	
	H. melancholicus	Bermudagrass	
	H. microlobus	Kentucky bluegrass	
	H. nannus	Bermudagrass Zoysiagrass	
	H. platyurus	Kentucky bluegrass	
	H. pumilus	Kentucky bluegrass	
Sting nematode[100,116,117]	*Belonolaimus gracilis*	St. Augustinegrass	Yellowing of turf foliage may be evident; stunting of root systems with lesions evident.
	B. longicaudatus	Bermudagrass St. Augustinegrass Zoysiagrass	
Stubby root nematode[116,117] (Figure 5.13)	*Trichodorus christiei*	Bermudagrass Tall fescue Kentucky bluegrass St. Augustinegrass	Stunting and yellowing of top growth; large brown lesions on roots; root tip swollen due to feeding.
	T. primitivus	Bermudagrass	
	T. proximus	St. Augustinegrass	
	Trichodorus spp.	Centipedegrass Zoysiagrass	
Stunt nematode[91,92,114]	*Tylenchorhynchus actus*	Bermudagrass	Stunting and thinning of top growth; short stunted root system; brown discoloration of roots; lesions usually lacking; wilting often evident.
	T. claytoni	Annual bluegrass Kentucky bluegrass Creeping bentgrass Zoysiagrass	
	T. dubius	Creeping bentgrass Kentucky bluegrass Fine-leaf fescue Annual bluegrass	

Table 5.1. Continued

Common Name	Scientific Name	Associated Turfgrass Species	Symptoms
Ectoparasitic Nematodes (continued)			
	T. maximus	Kentucky bluegrass	
	T. nudus	Kentucky bluegrass	
	Tylenchorhynchus spp.	Centipedegrass St. Augustinegrass Zoysiagrass	
Needle Nematode	*Longidorus* spp.	Creeping bentgrass	Stunting of top growth; reduction of root system.

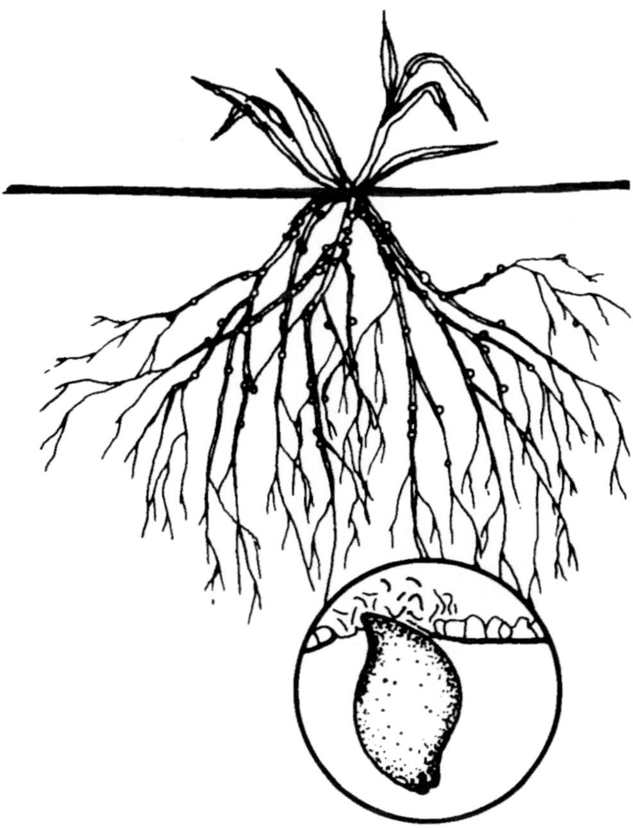

Figure 5.10. Turfgrass roots infected with cyst nematodes.

Figure 5.11. Cyst nematode damage on a Kentucky bluegrass turf.

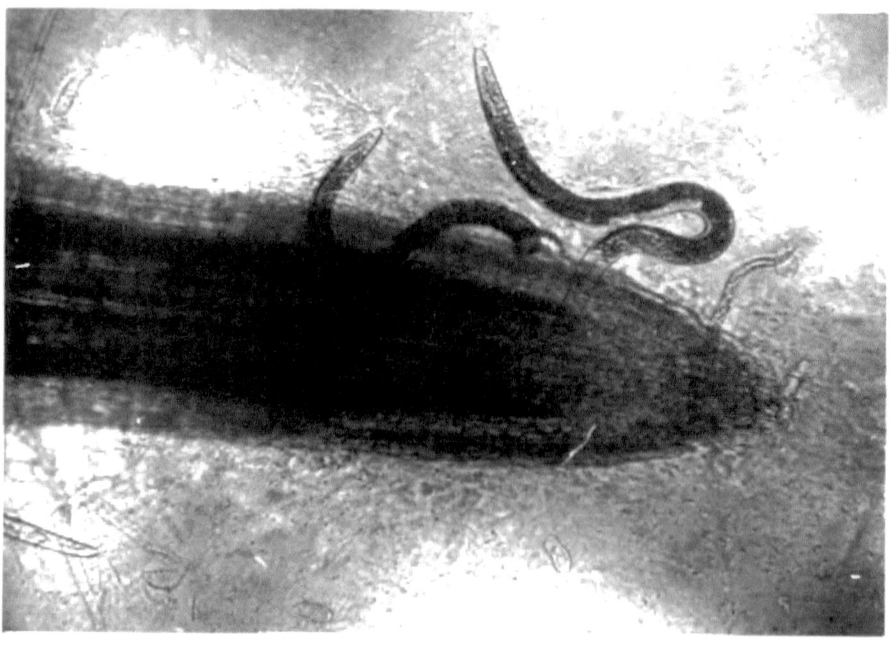

Figure 5.12. *Trichodorus* nematode feeding on root tip.

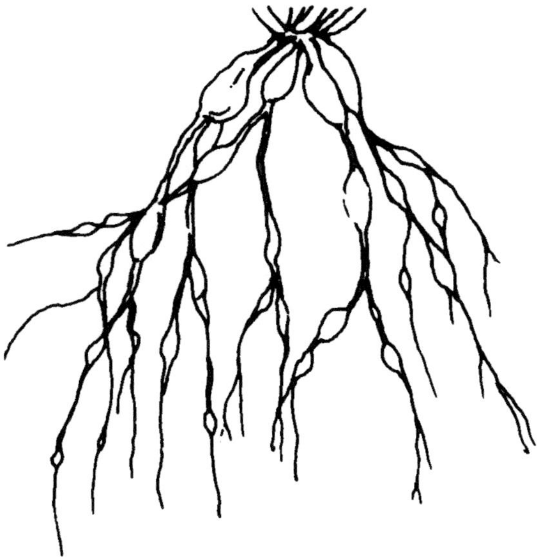

Figure 5.13. Root-knot nematode galls on a turfgrass root system.

MANAGEMENT

Sanitation

Sanitation is important with most diseases, but it is especially important in preventing nematode problems. Clean seed or sod is of little use in preventing foliar fungal diseases because most are airborne. It is different with nematodes. They have to be in the soil before it is seeded or sodded, brought in on the seed or sod, or brought in later with topdressing soil or cultivating equipment. Starting with nematode-free soil is a good way to prevent or delay severe nematode problems. If you are buying vegetative plant parts, such as sod, stolons, or sprigs for establishment, and the plant parts come from a nursery infested with nematodes that are pathogenic to turfgrass, chances are that you will acquire a nematode problem. The same is true of topdressing material. Having a nematode analysis before you purchase vegetative parts or topdressing may be well worth the minimal cost.

Fumigation prior to planting or replanting is advisable for soil already infested with nematodes. Fumigation of a golf course from tee to green is expensive, but it is within most budgets, at least for the greens. Even if the surrounding areas are infested, at least the young turf will have a chance to establish itself without the added stress of nematodes. It may be several years before the nematodes reinvade the fumigated area and become sufficiently numerous to create a problem.

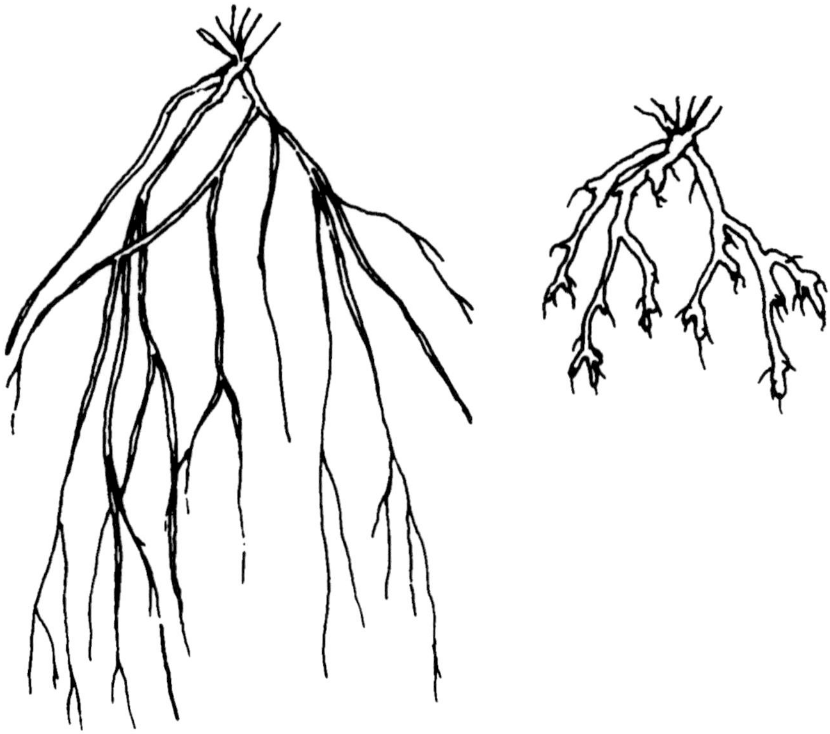

Figure 5.14. Short, swollen roots of a turfgrass plant attacked by stubby-root nematodes (right); normal, healthy grass root (left).

But do not go to the trouble of fumigating and then bring in nematode-infested plants or soil.

Cultural Practices

Watering

The development of nematode symptoms can often be prevented by adequate and properly timed watering. Frequent light watering is better than a few deep waterings. Although deep watering is usually recommended to develop a good root system, this works only on healthy turf. A nematode-infected turf is not healthy. It has a shortened, devitalized root system, with most of the roots in the top inch of soil or, even worse, in the top inch of thatch. When the top inch dries out (the thatch will dry out faster than the soil), the plants begin to wilt and die. Any water below the top inch is useless. The old theory that the plant roots will follow the water down does not apply, because the nematodes will prevent this from happening. During periods of hot weather, light syringing around noon may be necessary to prevent wilting.

Fertilization

In cooler regions of the world, proper fertilizing, like proper watering, can alleviate a nematode problem so satisfactory turf can be maintained. This is not true in the warmer climates. It is usually wise to keep phosphorus and potassium at adequate levels based on soil tests (see Chapter 7). Nitrogen should be applied lightly (less than one-half pound of actual nitrogen per 1000 square feet [0.25 kg per 100 square meters]). High levels of nitrogen increase top growth, which is perhaps already too great for the nematode-damaged root system, and will produce succulent plants, which are more susceptible to heat and drought stress. Moreover, a devitalized root system cannot use large amounts of nitrogen efficiently, and the excess is lost through leaching or volatilization.

Heald and Burton[65] found that plots fertilized with organic nitrogen (activated sewage sludge, or Milorganite) had a smaller population of *Belonolaimus longicaudatus* than plots receiving inorganic nitrogen (ammonium nitrate). Fertilization with organic nitrogen also resulted in increased root and top-growth weights.

Chemical Management

There are two types of nematicides, fumigants and nonfumigants. Fumigant nematicides usually give broad nematode control,[10] but they must be applied to the soil prior to planting. Nonfumigant nematicides that can be applied to an infected turf tend to be effective against some species but not against others, so that although the dominant nematode species may be reduced, other species that previously were not a problem may increase to the point that they become one.[90,93] Both fumigant and nonfumigant nematicides have a drawback common to all soil pesticides—their effectiveness depends somewhat on the type and texture of the soil. They are most effective on light sandy soils that are low in organic matter, and are least effective in heavy clay, silt-type soil, and soils high in organic matter, such as muck and peat soils. Higher rates of application may therefore be required in soils that are heavy or that contain significant amounts of organic matter.

It seems that the timing of nematicide treatments is important. If it is true that nematodes weaken a grass plant by shortening its root system, or that they open the way to fungal attacks by altering the plant's physiology, then nematicides should be applied early, to allow recovery of weakened roots and to prevent fungal infection. On the other hand, if nematodes simply put additional stress on plants already weakened by disease, then nematicides should be applied later in the season (Figure 5.15). There is probably no single correct answer.

Fumigant Nematicides

Fumigants include methyl bromide, chloropicrin, and 1,3-dichloropropane alone or in combination with chloropicrin and metham sodium. As mentioned earlier, they must be applied prior to planting. They should be applied at a depth of 6 to 8 inches (15 to 20 cm) (the depth of a normal turfgrass root zone) and allowed to volatilize and move up and out of the soil, killing the nematodes in the process. It is ideal if you can use specialized fumigation equipment that delivers the fumigant at 6 to 8 inches (15 to 20 cm) below the surface, but this is often not possible.

The labels on most fumigants specify an average number of days that must elapse between treatment and planting or replanting. However, if the soil temperature remains cool or if rainy weather keeps the soil excessively wet, it may take longer for the fumigant to volatilize, and you must postpone planting. The soil temperature must be at least 50°F (10°C) for fumigant nematicides to work. Lower temperatures may result in poor control or delayed planting or both.

Figure 5.15. Turf wilting in check due to high ring and stunt nematode populations. No wilt symptoms evident in nematicide-treated plot.

Nonfumigant Nematicides

Nonfumigant nematicides are organophosphate or carbamate derivatives, which are highly toxic to humans (see the LD_{50} chart in Chapter 6). They are usually sold as granular formulations, but this practice may be changing because birds pick up the granules.

The key to effective application of nonfumigant nematicides is getting the active ingredient into the root zone. Heavy watering following a nematicide application is recommended, and aerification before application will help to get the nematicide down to the root zone.[91]

Users often associate nematicides with fungicides rather than with insecticides, a habit of thought that may have dire consequences. All pesticides should be handled with extreme caution. Most people, however, tend to get a little careless when handling fungicides, especially when previous carelessness has caused no noticeable harmful effects. Careless handling of nematicides can be fatal! (See Chapter 6 for information on the safe use of pesticides.)

Resistant Cultivars

Little research has been done on the resistance of turfgrass cultivars to nematodes. Although it is important to study turfgrass resistance to all nematodes, priorities need to be established. Turfgrass scientists should not hope for a supercultivar resistant to all nematodes any more than they should expect to find one cultivar that is resistant to all diseases or to all insects. The one or two most severe nematode problems on a given species should be determined, and then a program should be established to breed for resistance to those nematodes. Minor nematode problems can, it is hoped, be solved culturally, the theory being that, like minor fungus problems, they should be easier to manage culturally and, if need be, chemically. Of course, a major assumption is that the new cultivar will not be susceptible to new and highly pathogenic nematodes. Significant research will need to be done over the next 10 to 20 years if we are to understand nematode problems in turf. Important topics for research include nematode-fungal interactions, resistant cultivars, chemical management, and biological management.

THE BLACK LAYER

Introduction

Sand topdressing is an excellent cultural practice. It has allowed many greens that were in poor playing condition because of poor soil structure to be converted to excellent greens without going through the drastic process of reconstruction. However, it was a classic example of a cultural practice being implemented without the necessary scientific research being done first.

This led to superintendents discovering problems associated with sand topdressing as they went along, rather than having the benefit of research that could have prevented many of these problems.

Some of the problems that have been associated with sand topdressing are localized dry spots, nutrient deficiencies, bacterial wilt of bentgrass greens, and the black layer.

The Role of Sulfur

Black layer symptoms occurs when sulfur or sulfur derivatives, such as sulfate, are reduced to the sulfide in the form of hydrogen sulfide, which is an unstable gas. The sulfide from hydrogen sulfide combines readily with metal ions in the soil like Fe, Mn, and Mg to form the corresponding sulfide (i.e., iron sulfide). The metal ions form a dark precipitate; hence the black layer.

Black layer is nothing new and has existed in many areas of the country for years, especially where high sulfate content irrigation water was used or where sulfur was used for pH reduction. The epidemic outbreak that occurred in the mid '80s appears to be directly related to use of calcareous sands for topdressing. These sands, because of their calcareous nature, had a high pH and superintendents assumed they had to use sulfur to reduce the pH in order to grow turf.

Initially, it was assumed that the black layer was something unique to sand greens or sand topdressed greens. However, black layers have been found in soil greens, tees, fairways, and roughs. The only thing they had in common was the presence of high levels of sulfur in the soil.

Symptoms

Initial symptoms are a general thinning of the turf, particularly in the low areas of the greens. If conditions for black layer continue, the grass will eventually turn brown as the plants die. If the soil is examined, a black layer ranging in size from an inch to several inches will be observed. The location of the layer will vary from near the surface to several inches down in the soil. The "black layer" can also occur as pockets of blackness rather than a complete layer, and as an irregular black mottling of the roots. If the black layer is smelled, a foul odor will be detected. It is very similar to rotten eggs.

Cause

Berndt et al.[8] showed the black layer results from having excess sulfur in the system and then having the system become anaerobic. When this occurs, sulfate is reduced by sulfate-reducing bacteria to H_2S.[9] The H_2S is a gas that is toxic to the root system of the grass plant and can cause the plants to die. It is a highly unstable gas that combines readily in the soil with cations of covalent plus 2, such as Fe, Mn, and Mg, to form the corresponding sulfide, i.e., iron sulfide. It is the iron sulfide that gives the black color to the layer. The tying up of the hydrogen sulfide by ions in the soil is beneficial, since it ties up the toxic H_2S. However, in doing so, it forms a black precipitate that interferes with the downward movement of water, further perpetuating the anaerobic condition.

Excess sulfur in the soil usually comes from three sources: elemental sulfur which is applied for pH reduction, or from sulfate in irrigation water, or where it is added to the irrigation water as sulfuric acid for pH reduction. An additional source can be from fertilizer containing various forms of sulfate, like potassium sulfate.

Anaerobic conditions occur when oxygen is eliminated from the soil. This is usually caused by excess irrigation, traffic (causing compaction), soil layering or heavy or prolonged rainfall. Whereas irrigation can be managed, little can be done to eliminate traffic on greens, and nothing can be done to prevent the rain from falling on a golf course. Furthermore, as sulfur is oxidized to sulfate under aerobic conditions, it becomes an oxygen sink by removing what little oxygen remains after the plant roots have used their share for nutrient and water uptake, and the microorganisms have used their share to respire. This reduces the oxygen level further, directly contributing to the anaerobic condition.

Is it necessary to oxidize sulfur to sulfate? How else is pH reduction going to occur? It is the release of hydrogen ions during the oxidation of sulfur that results in the reduction of the pH. In order for the oxidation of sulfur to occur, oxygen must be taken from the soil. Since there is already precious little oxygen present in the soil for use by the plant in the uptake of water and nutrients, and for use by the microorganisms to carry out their life functions, when the sulfur utilizes the remaining oxygen in the soil as it is being converted to SO_4, the soil then becomes anaerobic. Thus, people who say sulfur does not directly contribute to the anaerobic condition of the soil simply don't understand soil chemistry.

Myths of Sulfur and pH Reduction

Many people mistakenly believe that in order to maintain healthy turf the pH has to be between 5.5 and 7.0, in spite of the fact that they have been successfully growing turf at a higher pH for many years. This is based on facts learned in basic soils classes or at some turf conference. Unfortunately, this information is not based on turf research, but on research done on corn and wheat. There are good reasons to have the pH between 5.5 and 7.0 for corn and wheat production. The major nutrient that is lacking at a high pH is phosphorus, which is essential for the formation of the wheat heads and the corn ears. It is also difficult to run a fertilizer spreader down a corn field when the ears are being produced and the corn is 6 feet (2 meters) high. A golf course superintendent does not have that problem. He can run a fertilizer spreader over his turf every day, if necessary, to apply nutrients that are unavailable at a high pH, and it is a lot safer than trying to lower the pH with sulfur, as you shall see shortly.

Many greens that are topdressed with calcareous sands contain at least 20% $CaCO_3$.[17] The upper three inches (7.6 cm) would contain approximately 5,000 lb (2270 kg) of $CaCO_3$. It is estimated that if the normal recommended rate for pH reduction of 5 lb of sulfur per 1000 sq ft (2.4 kg per 100 square meters) were applied, it would take approximately 1000 years to neutralize the $CaCO_3$ in such a green.[18] This of course assumes that you are not irrigating from a calcium or sodium aquifer which could be applying the equivalent of 10 to 20 lb (4.5 to 9 kg) of Ca or Na per season.

Another popular approach today is to inject sulfuric acid into the irrigation systems. It has been calculated that it would take 34 tons of sulfuric acid per acre (76 metric tons per hectare) to neutralize 1% of the $CaCO_3$ in the top 3 inches (7.6 cm). At $1.45 per pound (0.5 kg), this would cost approximately 100,000 dollars![18] Save your money and your job. Forget about pH reduction with sulfur. Add the nutrients as they are needed; it is a lot cheaper and a lot safer.

It all reminds me of consulting in the Caribbean. The bermudagrass greens had little fertilizer burn marks on them from using ammonium sulfate. When I inquired as to why he was using ammonium sulfate, he explained it was for pH reduction. I asked him how long he had been using ammonium sulfate, he said, "ten years." I asked him what the pH was then, he said "8.4." I asked him what the pH was now, he said "8.4." I suggested that should tell him something, since the pH hadn't gone down in ten years. To which he replied, "but it hasn't gone any higher." And it won't! The pH of his soil was the same as the pH of his irrigation water . . . as it is on most golf courses, and it is impossible to lower the pH under such conditions without injuring the turf.

MANAGEMENT

The simplest way to manage the black layer is to avoid applying sulfur for pH reduction and to limit the use of fertilizers that contain sulfur products. Where the black layer occurs from high sulfate content water, placing the greens on city or town water is advised. If this is not possible, irrigate lightly using syringe cycles, avoid deep irrigation cycles completely, and do multiple aerifications throughout the summer months.

If the soil has become anaerobic, getting oxygen into the soil is important in stopping the development of the black layer. This can be accomplished by opening up the soil through aerification. Oxygen can also be put in the soil through the use of nitrate fertilizers like potassium nitrate. By oxidizing the soil, the severity of the black layer should be reduced. Extreme caution should be used when applying nitrate fertilizers to the greens because of their burn potential. Don't apply any more than one-fourth lb (.12 kg) of actual nitrogen per 1000 sq ft (100 square meters) per treatment.

CHAPTER 6

Fungicides

Fungicides belong to a group of chemicals called pesticides, which are used to protect plants from pests. Other pesticides include insecticides, which control insects; nematicides, which control nematodes; and herbicides, which control weeds. Too often the term pesticide is used incorrectly to mean only insecticides. Actually, it refers to the entire group of chemicals used to control plant and animal pests.

The term, fungicide, is applied to all chemical plant protectants that manage fungal diseases. Many fungicides, however, do not kill the fungus, but merely inhibit its growth by preventing germination of spores or by stopping growth of the hyphae or mycelium after germination. Such chemicals are really fungistats (fungal growth inhibitors) rather than fungal killers. The benzimidazole fungicides (benomyl and thiophanate methyl,) and *D*e*M*ethylation *I*nhibitor fungicides (DMI) (fenarimol, propiconazole, and triadimefon) are examples of fungistats. Even though they are fungistats, when applied as eradicants (i.e., after the disease is present), they will stop the spread of a fungus by preventing it from entering new plant tissue.

Table 6.1 Common Names and Trade Names of Widely Used Fungicides

Common Name	Trade Name
Anilazine	Dyrene, Dymec 50
Benomyl	Tersan 1991
Chloroneb	Termec SP, Proturf Fungicide V
Chlorothalonil	Daconil 2787
Etridiazol (ethazol)	Koban, Terrazole
Fenarimol	Rubigan
Flutolanil	ProStar
Fosetyl-Al	Aliette
Iprodione	Chipco 26019, Rovral, Scotts Fungicide X
Maneb	Manazate 200
Mancozeb	Fore, Formec 80
Metalaxyl	Subdue, Proturf Pythium Control
Mercury compounds	
Inorganic	Calo Clor, Calo Gran
PCNB	Terraclor 75, Scotts F + F II, Turfcide II, Pennstar
PMA (PMAS)	PMA, PMAS
Propamocarb	Banol
Propiconazole	Banner
Thiophanate methyl	Fungo 85, Spot Kleen, Proturf Systemic Fungicide, Topmec 70, Clearys 3336
Thiram	Tersan 75, Spotrete, Thirmad
Triadimefon	Bayleton, Scotts Fungicide VII
Vinclozolin	Vorlan, Touche, Curalan

GENERIC NAMES, CHEMICAL FORMULAS, AND TRADE NAMES

Most turfgrass managers know the fungicides they use by their trade names. Trade names are given to fungicides by manufacturers or distributors; the names are registered with the U.S. patent office to prevent others from using them for their products. This practice is intended to protect the company selling the fungicide, but it also protects the user. It is confusing enough to try to remember the trade names of the many fungicides; it would be even harder to memorize the chemical compositions of different fungicides with the same trade name. The fungicides are listed by generic names and trade names in Table 6.1 (the listing is alphabetical by generic name).

Some fungicides are sold as combination products containing two or more fungicides. Some of these are listed in Table 6.2.

Is a fungicide sold by one company better than the same fungicide sold by another company? The answer is no. As a matter of fact, most of the time a fungicide is manufactured by one company and sold to another for distribution or packaging under different labels or trade names. The only possible substantive difference between the two products is a difference in the concentration of the active ingredient (Figures 6.1 and 6.2).

Table 6.2. Composition of Some Fungicide Mixtures

Generic Makeup of Mixture	Trade Name
Fenarimol + chlorothalonil	Twosome
PMA + thiram	Proturf Broad Spectrum Fungicide
Thiophanate methyl + thiram	Bromosan
Thiophanate methyl + maneb	Duosan
Metalaxyl + maneb	Pace
Thiophanate methyl + iprodione	Scotts Fluid Fungicide
Triadimefon + metalaxyl	Scotts Fluid Fungicide II
Triadimefon + thiram	Scotts Fluid Fungicide III

Figure 6.1. MBC, the active ingredient in benomyl and thiophanate methyl, not controlling resistant dollar spot compared to Daconil 2787 (chlorothalonil).

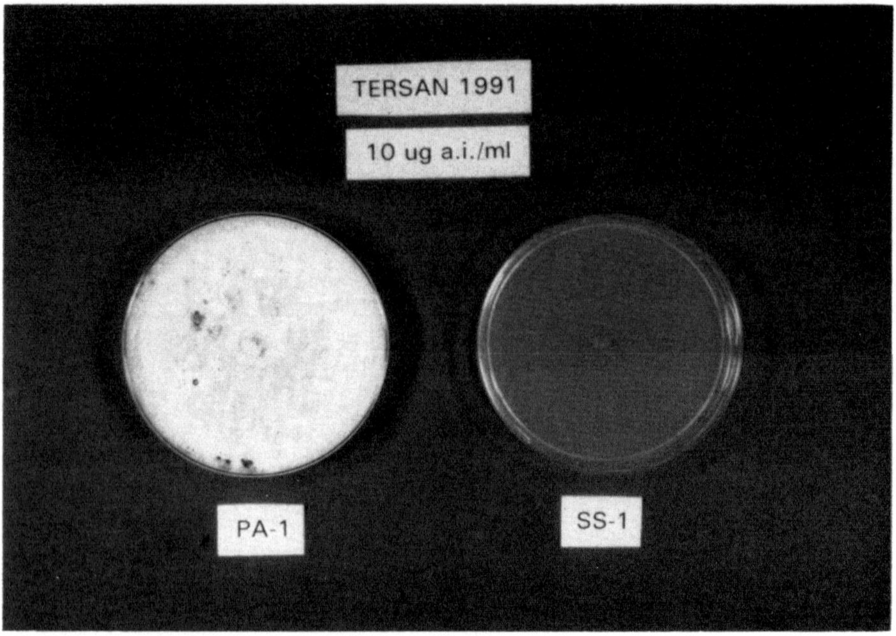

Figure 6.2. Tersan 1991 not inhibiting the growth of the PA-1 (benomyl-resistant) strain and completely inhibiting the growth of the SS-1 (wild type) strain of *Sclerotinia homoeocarpa*.

SYSTEMIC AND NONSYSTEMIC (CONTACT OR SURFACE) FUNGICIDES

The major difference between systemic and nonsystemic (contact) fungicides (Table 6.3) is the means by which they control turfgrass diseases. A contact fungicide is applied to the foliage, where it prevents or halts infection. It forms a protective shield around the outside of the plant. The contact fungicides have the general disadvantage of being exposed to weathering and photodecomposition, and on turf they are exposed to an additional hazard: mowing. Turfgrass is constantly being mowed, and as the old foliage is mowed off, the new foliage that replaces it is unprotected. This is especially a problem on greens that are mowed daily. Also, turf is watered nightly during most of the summer, and so the fungicide that isn't mowed away may be washed away.

The systemic fungicides differ from contact fungicides in that while they also provide an external barrier to infection, they are in addition absorbed by the plant, and translocated throughout. The benzimidazole fungicides are only translocated upward (acropetal). The dicarbox-

Table 6.3. Fungicides

Contacts	
	Triazines—anilazine
	Substituted aromatic compounds—chlorothalonil, PCNB, chloroneb
	Heavy metals—mercuries
	Ethazol (Thiazoles)—terrazole
	Dithiocarbamates—maneb, mancozeb
Systemic	
	Demethylation Inhibitor (DMI)—fenarimol, propiconazole, triadimefon
	Dicarboximides—iprodione, vinclozolin
	Benzimidazoles—benomyl, thiophanate methyl
	Acylalanines—metalaxyl
	Carbamates—propamocarb
	Aluminum ethyl phosphate—fosetyl-Al

imides, DMI, metalaxyl and fosetyl-Al fungicides appear to be absorbed by the plants and translocated in both directions in the plant (acropetal and basipetal), since they manage root diseases when applied to the foliage. The dicarboximides are often listed as contact fungicides, but research by Danneberger and Vargas[29] has shown them to be systemic. In addition, there is no way a contact fungicide could give 3- to 4-week control of dollar spot on golf greens that are mowed every day, as the dicarboximides do, and not be systemic.

Since the systemic fungicides can be absorbed, they can persist after the treated foliage is mowed off; new foliage will obtain some systemic fungicide from redistribution and from the fungicide that the roots continue to pick up from the soil. The systemic fungicide will eventually be tied up, broken down (degraded), or leached from the soil, and will then no longer be available. At such time, an additional application will be needed. A systemic fungicide will give protection for 3 to 4 weeks, depending upon the soil type and disease pressure. Turfgrass that is growing in sandy soil usually requires more frequent fungicide application than does turfgrass growing in loam, silt, or clay.

Systemic fungicides have controlled diseases, such as necrotic ring spot, summer patch, and stripe smut that once were considered uncontrollable. Necrotic ring spot and summer patch are primarily root and crown diseases, and management requires a systemic fungicide that can be taken up by the roots, or translocated to the roots. It would be impossible to completely surround the root and crowns of an established turfgrass plant with a contact fungicide without removing the plant and dipping it into a fungicide solution. Stripe smut is a systemic disease which is internal to the turfgrass plant, and external fungicides will have no effect on it. (See Stripe Smut in Chapter 4.)

The systemic fungicides are not the panacea they were once thought to be. Their greatest assets have been their low-concentration residual management of dollar spot and the fact that they are the only fungicides that manage necrotic ring spot, summer patch, and stripe smut. However, turfgrass pathogens like *Sclerotinia homoeocarpa*, the cause of dollar spot, and *Pythium* spp. that causes *Pythium* blight, have developed resistance to some of them.

RESISTANCE TO FUNGICIDES

Resistance occurs when a fungus which was sensitive to a fungicide becomes resistant to it. Resistance to fungicides is not new; its occurrence has been known for years.[86] It is also no stranger to the turf area, as cadmium and anilazine-resistant strains of *Sclerotinia homoeocarpa* have been reported by Cole, Taylor, and Duich.[23] However, resistance to contact fungicides has never been as widespread and developed as quickly as has resistance to the systemic fungicides. Resistance to the benzimidazole systemic fungicides, for example, usually occurs during the second or third season that these fungicides are used. The speed with which resistance has developed to the benzimidazole fungicides is related to their specific mode of action. They inhibit cell division by preventing the sister chromatids from migrating to the poles during the metaphase of mitosis.[37] This means that fungi that can prevent the fungicide from doing this will not be controlled by it, while the sensitive fungi will. It would be more difficult for a fungal strain to block the action of a contact fungicide, since such fungicides are believed to have many sites of action, and they affect many processes of the fungal cell.

Introduction

Truly one of the most misunderstood areas in all of turfgrass disease management is resistance to turfgrass fungicides. This includes the role that the mode of action of the fungicide plays, the role the various types of pathogens play, and most certainly the outmoded ways used delay the resistance. All of these factors will be discussed in the following pages. Hopefully, when you are through with this chapter, you will have a better understanding of resistance; why it has occurred in the past; and how to delay it from occurring in the future.

History

Prior to 1969, resistance to fungicides was primarily confined to cadmium resistance in dollar spot caused by *Sclerotinia homoeocarpa*.

Resistance to the cadmium fungicides occurred only after many years of use, but this was not considered a major problem since resistance was not widespread and resistance to the other contact type of fungicides had not occurred. The development of the systemic fungicide in the late 1960s changed all that forever. The first reported resistance to a fungicide by a turfgrass pathogen was to *Erysiphe graminis*, the cause of powdery mildew, by the benzimidazole fungicide benomyl (Tersan 1991).[153] Since this was such a minor disease, the research was ignored or dismissed with the attitude that benomyl wasn't all that effective against powdery mildew, anyway. The following season reports began to appear that benomyl was no longer controlling dollar spot.[23] Resistance could no longer be ignored, as it was occurring on a major turfgrass disease and all over the country.

Since we had no previous experience in dealing with widespread resistance to a major fungicide, it's understandable that some simple solutions were suggested to solve the problem. It is unfortunate, though, that many of those myths still exist today. One solution offered was to use a contact fungicide for a couple of sprays to eliminate the new resistant strain. Another was not to use benomyl for one year, and the resistant strain would disappear. Neither solution worked. The contact fungicides never eliminated or altered the wild type strain of *S. homoeocarpa*; why should they eliminate the resistant strains? The other solution of not using benomyl for a season assumed the benzimidazole-resistant strain would disappear from the population because it would lack fitness or pathogenic aggressiveness to compete with the wild type strain of *S. homoeocarpa*. This was more logical, but no more valid. The benzimidazole-resistant strains had excellent fitness and pathogenic aggressiveness, and consequently persisted. In fact, it is difficult to find a golf course in the U.S. where it is not the dominant strain. It is so dominant and widespread that if we could all leave the planet for a hundred years, come back, and rediscover benomyl, upon field-testing it against dollar spot, our conclusions would be that benomyl is not an effective fungicide for the control of dollar spot.

Terminology

Wild type sensitive strain: the original or naturally occurring strain of the fungus that existed in the turf before selective fungicides were applied. This strain is sensitive to the fungicide, and controlled by it.

Resistant strain: a strain of the fungus which is not controlled by a fungicide that is effective in managing the original wild type strain.

Single-site fungicides (SS): (systemic fungicides) fungicides that keep fungi from growing by attacking one or a few vital systems of a fungus, usually controlled by a single gene in the fungus. All these fungicides also happen to be systemic fungicides. The single site of action of metalaxyl and the benzimidazoles are controlled by a single gene in the fungi. The primary site of action of the DMI fungicides is multigenic.

Multi-site fungicides (MS): (contact fungicides) fungicides that control the fungus by attacking several different vital systems of a fungus. These fungicides also happen to be contact fungicides.

Fungicides: The older contact fungicide, or multi-site specific fungicides (MS) act on many vital systems of the fungus. The literature is full of papers trying to determine the exact site of action of these fungicides on fungi. Or, more simply stated, how does the fungicide control the fungus? Most research showed that there was not one specific site of action, but several, affecting many vital systems of a fungus. This being the case, a fungus would have to have natural mutations occur for the numerous genes that control these many sites of action in order for the fungus to become resistant to an MS fungicide. This is highly unlikely, as evidenced by the almost complete lack of resistance developing to these fungicides.

Systemic or single site (SS) fungicides primarily attack one vital system of a fungus. If this single site is controlled by a single gene, as is the case for metalaxyl and the benzimidazoles, a mutation of a single gene is all that is necessary to give a fungus resistance to these fungicides. So, as you can see, it is easier for a fungus to become resistant to an SS fungicide than to an MS fungicide. It is important to remember that neither SS nor MS fungicides cause mutations; they simply apply selective pressure to a fungal population that allows the resistant strain to become dominant as the wild type sensitive strain is eliminated from the population.

FUNGICIDE FACTOR

Resistance to Benomyl and Cross Resistance to Thiophanate Methyl

The active breakdown product for both benomyl and thiophanate methyl fungicides is a compound called methyl benzimidazole carbamate (MBC). The fact that the original chemical structures of benomyl and thiophanate methyl do not look alike caused some confusion in the field. Even though laboratory experiments and university field trials had shown that strains of *S. homoeocarpa* that were resistant to benomyl were also cross-resistant to thiophanate methyl, it was still sold as having a different mode of action that would control the benomyl-resistant strain of *S. homoeocarpa*. It didn't—as many turf managers that used it against the benomyl-resistant strains found out.

Cross-resistance among the dicarboximides and the DMI fungicides also exists. We have found that strains of *S. homoeocarpa* resistant to iprodione were also cross-resistant to vinclozolin. Likewise, *S. homoeocarpa* strains resistant to one of the DMI fungicides, like triadimefon, were found to be resistant to the other DMI fungicides, like fenarimol and propiconazole.

Factors Involved in Resistance

The two major factors involved in resistance are the fungus and the fungicide. While you may be thinking this is logical, much of what has been written about resistance deals primarily with the fungicides. Little has been written about the role the different types of fungi play in the development of resistance.

Effects of SS Fungicides on Fungal Population

The efficacy studies with the benzimidazole fungicides show long-term control of dollar spot with single high-rate applications. For example, several researchers showed season-long control of dollar spot with 2 to 8 oz/1000 sq ft (63.5 to 254.4 mL per 100 sq meters) applications on their research plots. This was assumed to be due to the fungicide remaining in the soil and being continually taken up by the turfgrass plant over the year. The same long-term control, however, was not achieved on golf course greens. The question was, "Why the difference?" To answer the question, we began to inoculate the benzimidazole fungicide plots with clippings infected with dollar spot. To our surprise, the plots immediately became infected. What we were observing was not season-long control of dollar spot from the residue of the fungicide in the soil, but season-long control because the fungicide had *eliminated* all the sensitive, wild-type strains of *S. homoeocarpa*. Under golf course conditions, golfers were tracking dollar spot from the fairways and roughs back onto the greens. Under the research plot conditions, only the mowers could spread the disease, and based on the observed results, this wasn't readily occurring. Since *S. homoeocarpa* does not produce spores that can be carried by the wind to infect or reinfect turf areas, its means of spread is mechanical, by people or machines. What does all this have to do with resistance?

SS fungicides eliminate only sensitive strains from the fungal population, whereas MS fungicides suppress the entire fungal population. Secondly, SS fungicides act on sensitive fungi after they have germinated. The benzimidazole fungicides, for example, inhibit cell division. If the fungus can't divide and produce new cells, it dies when its stored energy is used up, and is decomposed by other microorganisms. The dicarboximide fungicides cause hyphal bursting (they also inhibit spore germination). The germ tube from the fungal spore bursts, releasing the contents of the cell. When this happens, further growth ceases and the fungus dies. The DMI fungicides prevent the formation of sterols, preventing the fungus from developing cell membranes. Without cell membranes, cells cannot be made, and the fungus dies. But all of these processes take place after germination. The fungicides have no effect on the fungus until after it has germinated. The spore is much like a seed. Once a seed has germinated it must continue to grow and begin to photosynthesize and produce food before its energy is depleted. Otherwise, it will die because it cannot revert back to a seed after it has germinated. Likewise, a fungal spore, once germinated, must continue to grow and find a food source before its energy is depleted; otherwise, it will die.

Many of the MS fungicides prevent spore germination. This preserves the spore as a viable propagule that can later germinate and infect a plant, once the fungicides disappear. The SS fungicides, with the exception of the dicarboximides, all act after germination. The spores apparently don't recognize the presence of the SS fungicide, since preventing spore germination is not part of their mode of action. They germinate in the presence of the SS fungicide, and are then eliminated from the population.

Knowing that SS fungicides can eliminate sensitive strains tells us that higher rates and shorter intervals will eliminate wild type strains faster from a population than will lower rates and longer intervals. [A point to be discussed more later.]

Persistence of Resistant Strains

So far, we have discussed the resistance of *S. homoeocarpa* to the benzimidazole fungicides. Resistance of strains of *S. homoeocarpa* has been reported to other SS fungicide groups like the DMI fungicides, fenarimol, triadimefon, propiconazole, and the dicarboximide fungicides, iprodione and vinclozolin. Resistance to the dicarboximides has

Table 6.4 Degree of Resistance and Persistence Among the SS Fungicides

Fungicide Class	Degree of Resistance	Persistence
Benzimidazoles	High	Yes
Dicarboximides	High	No
Acylalanines	High	Yes
Organic phosphates	Low	No
DMIs	Low	Yes

also been reported in *F. nivale*.[17] The strains of *S. homoeocarpa* that were resistant to the dicarboximides also had resistance to either the benzimidazoles[35] or the benzimidazole and DMI fungicides as well.[163] A fungus with resistance to two or more completely unrelated fungicide groups is said to have double or multiresistance. *S. homoeocarpa* strains with double-resistance to either benomyl-DMI fungicides or resistance to the benzimidazole-dicarboximide-DMI fungicides have been found on golf courses. Isolations from the golf course with the double-resistant strains one year later, after dicarboximide fungicide use had been discontinued, revealed that the dicarboximide double-resistant strain had disappeared,[35] and only the benzimidazole-resistant and benzimidazole-DMI-resistant strain remained. Persistence of resistant strains is obviously important, and differs among the fungicide groups, as can be seen in Table 6.4. Apparently strains of *S. homoeocarpa* that are double or multiresistant to the dicarboximide fungicides do not persist in the turf once the use of that group of fungicides has been discontinued. It appears that resistance may be more apt to occur in strains that already have resistance to another fungicide group, and after the use of dicarboximide fungicide is discontinued, they tend to disappear from the population.

Strains of *Pythium aphanidermatum*, the cause of *Pythium* blight, have been reported to be resistant to the phenylamide fungicide, metalaxyl.[127] From a persistence standpoint, the metalaxyl-resistant strains have declined but remained in the population. The wild type strains of *P. aphanidermatum* have reappeared once the use of metalaxyl was discontinued, although not in numbers large enough to allow the reuse of metalaxyl. This is in contrast to what has happened to the sensitive strains of *S. homoeocarpa*, which have failed to reappear even after the use of the benzimidazole was discontinued. This may be explained by the fact that *P. aphanidermatum* is a facultative parasite which has allowed the remaining wild type sensitive strains to grow as saprophytes in the absence of metalaxyl, whereas *S. homoeocarpa*, being a facultative saprophyte, would not have that ability. In addition, *P. aphanidermatum* produces a very sturdy survival spore which allows it to persist for long periods in adverse environments, including the presence of fungicides.

Strains of *S. homoeocarpa* resistant to benzimidazole fungicides and strains of *P. aphanidermatum* resistant to metalaxyl tend to persist in the turf even after the use of the specific fungicide has been discontinued. Strains of *S. homoeocarpa* resistant to the dicarboximides plus the benzimidazoles, or the dicarboximide plus the benzimidazole and DMI fungicides, or to the dicarboximides plus the DMI fungicides, tend not to persist in the turf. Strains resistant to the DMI fungicides that are now occurring in turf appear to be persistent as well. Some of these strains of *S. homoeocarpa* with resistance to the DMI fungicides have also been found to be multiresistant to the benzimidazole and dicarboximide fungicides. All the strains that have been isolated with resistance to the DMI fungicides were also double-resistant to the benzimidazole fungicides. How persistent the DMI-benzimidazole double-resistant strain of *S. homoeocarpa* will be, only time will tell. However, they do have pathogenic aggressiveness, which would tend to suggest they will be persistent.

Pathogenic Aggressiveness

If a fungus is to survive and compete with the wild type, it must have aggressiveness. It was originally believed, as stated before, that these resistant strains would lack pathogenic aggressiveness, and therefore would not survive in the population once the selection pressure of the fungicide was removed. However, the strain of *S. homoeocarpa* that is resistant to the benzimidazoles tends to be very aggressive. The same is true of the strain of *P. aphanidermatum* that is resistant to metalaxyl. In many cases, the resistant strains of the two pathogens were even more aggressive than the wild sensitive strains they replaced. The dicarboximide-benzimidazole-resistant strains, on the other hand, lack pathogenic aggressiveness, which undoubtedly contributed to their inability to survive in the population.[35]

Degree of Resistance

Another factor is degree of resistance. If the strains are highly resistant, as is usually the case with fungal strains resistant to the benzimidazoles and metalaxyl, then even increasing the dosage rate won't control them. But if the strains have a low level of resistance, as is often the case with resistance to the DMI fungicides when it is initially detected, then by simply increasing the dosage level they can be controlled. The levels of resistance can be seen in Table 6.4.

The low level of resistance to the DMI fungicides by strains of *S. homoeocarpa* can be controlled by increasing the rates and/or shortening the interval. This, of course, will only lead to increased levels of resistance to the fungicides, to the point where rates will become so excessive that it will become impractical to use a DMI fungicide. This type of resistance will express itself as incomplete control. Dollar spot may not get any worse but will persist at a low level, or it may be seen as reduction in length of control from 21–28 days down to 10–14 days of control from a single application.

FUNGAL FACTOR

As you will notice, the discussion has primarily revolved around *S. homoeocarpa*, the cause of dollar spot. The benzimidazole fungicides have been used to control brown patch, caused by *Rhizoctonia solani*, since 1969, with no reported case of resistance to the fungicides by *R. solani*. These same fungicides have been used since 1975 to control *Colletotrichum graminicola*, the cause of anthracnose, since 1975, with only one verified case of resistance.[36] If the benzimidazole fungicides have been used to control these diseases on the same courses where resistant strains of *S. homoeocarpa* are present, why have resistant strains of *R. solani* or *C. graminicola* to these fungicides not also developed? In order to answer that question we need to have a basic lesson in plant pathology.

Forms of Parasitism

Three types of parasites exist, based on their life cycles. They are as follows:

Obligate parasite: an organism that obtains all of its nutrition from another living host.

Facultative parasite: an organism that spends most of its life as a saprophyte, but can, under ideal conditions, become a parasite.

Facultative saprophyte: an organism that is primarily a parasite, but can for short periods of time live as a saprophyte, usually while the host tissue is still recognizable, before it becomes general organic matter.

The fourth type of organism is not a parasite, and therefore not a plant pathogen. It is called an:

Obligate saprophyte: an organism that obtains all of its nutrition from dead or decaying organic matter.

Obligate parasites and facultative saprophytes, once eliminated from the population, have no means of repopulating the turf area other than when they are parasites. Fungicide sprays are applied when they are infecting the plant, which contributes further to the elimination of sensitive wild type individuals from the population. Facultative parasites, on the other hand, spend most of their lives living in the thatch and/or soil as saprophytes and only when conditions become ideal for disease development do they become parasites. So during periods of time when the disease is not present and no fungicide applications are being made, they have a chance to repopulate the turf area. The ability of this type of organism to survive and increase in treated areas means less chance of resistance developing. However, heavy rates and shorter intervals may eventually lead to resistance developing to the SS fungicides in facultative parasites.

Comparison of a Facultative Parasite and Facultative Saprophyte and Their Tendencies To Develop Resistance to an SS Fungicide.

Pythium Blight and *Phytophthora* Late Blight

Figure 6.3 illustrates the difference in the life cycle of these organisms and the effect that SS fungicides have on their populations. Table 6.5 lists the turf diseases that are caused by these 3 types of organisms. You will note that most of the cases where fungicide resistance has been widespread have occurred in fungi in the obligate parasites and facultative saprophytes categories. To further illustrate this point, let's use the example of *Pythium* and *Phytophthora*, two genera of fungi managed by the fungicide metalaxyl. During the first and second season of metalaxyl use on late blight of potatoes caused by *Phytophthora infestans*, resistance to the fungicide in the pathogen became widespread throughout Europe and South America. *P. infestans* is a facultative saprophyte, and once sensitive strains were eliminated from the population they had no chance to repopulate saprophytically. Metalaxyl, on the other hand, has been used to manage *Pythium* blight of turfgrass caused by *Pythium aphanidermatum* for 7 years on over 8,000 golf

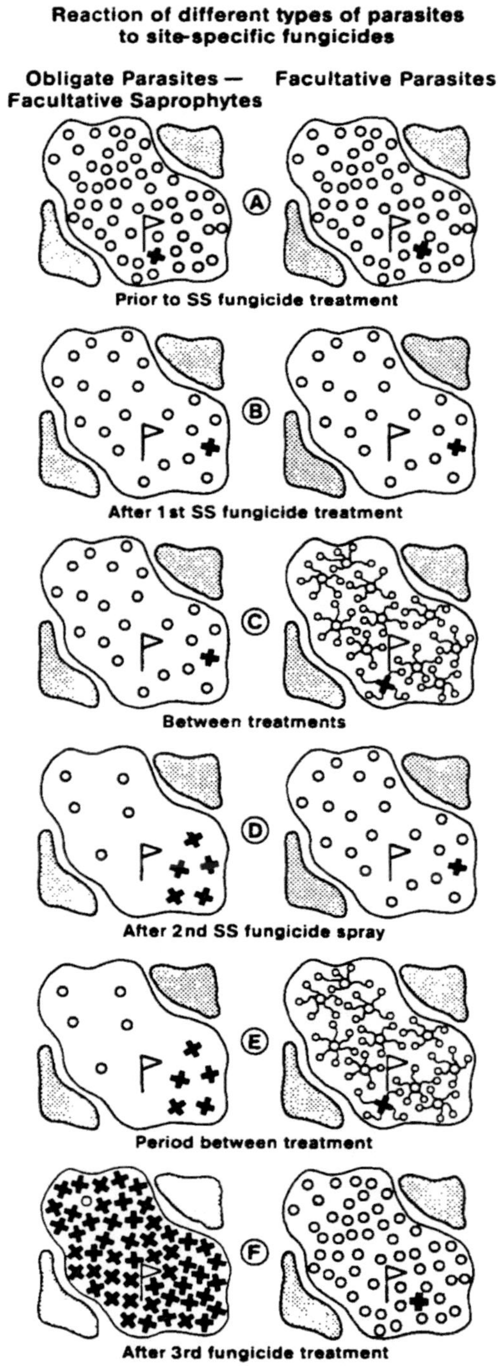

Figure 6.3. Reaction of different types of parasites to site-specific fungicides.

Table 6.5. Disease Classification Based on the Type of Parasitism Exhibited by the Fungus

Obligate **Parasites**	Facultative **Saprophytes**	Facultative **Parasites**
Rusts[a]	Dollar Spot[a]	Brown Patch
Powdery Mildew[a]	*Microdochium* Patch[a]	Anthracnose
Yellow Tuft	*Phytophthora*[a]	*Pythium* Blight
	Stripe Smut	

[a]Those diseases to which widespread resistance to SS fungicides has occurred, whether in turf or another crop.

courses, with fewer than 30 verified cases of metalaxyl resistance having been reported. One plausible explanation is the fact that *P. aphanidermatum* is a facultative parasite that has the ability to repopulate areas where the sensitive wild type strains have been eliminated by growing as a saprophyte when metalaxyl is not present. Documented cases of metalaxyl resistance were associated with heavy rates and short intervals being used. If lack of control or shorter intervals of control start occurring, the use of metalaxyl should be discontinued. Further use may completely eliminate the wild type sensitive strains of *P. aphanidermatum* from a golf course.

Dollar Spot, Brown Patch, and Anthracnose

A similar situation has occurred on golf courses where benzimidazole fungicides have been used for many years. The facultative saprophyte *S. homoeocarpa* (dollar spot) has developed resistance to these fungicides, whereas the facultative parasites *R. solani* (brown patch) and *C. graminicola* (anthracnose) have not. This may be because the facultative parasites spend most of their lives growing in the thatch and soil as saprophytes, and can repopulate areas where their numbers have been reduced by SS fungicides.

Another possibility that could explain the difference among fungi to develop resistance to SS fungicides includes the number of genes controlling single site mode of action. Whatever factors are involved, it is obvious that among fungal species there are differences in their tendency to develop resistance to the SS fungicides.

Microdochium Patch

The other major turfgrass pathogen that has developed resistance to benzimidazole and dicarboximide fungicides is *Microdochium* patch, caused by *Microdochium nivale*. *M. nivale* is also a facultative saprophyte. It is interesting to note the only report of resistance to *M. nivale* has come from the coastal areas of the Pacific Northwest, where the disease occurs from late August into June.[16] Under these conditions, multiple fungicide applications are required throughout the 9–10 month period. In other areas of the U.S. where *Microdochium* patch occurs less frequently and the benzimidazole fungicides are normally only used for spring and fall control of *Microdochium* patch (mercury and PCNB fungicides are used for winter control) and then only when symptoms occur, and not as part of a regular spray program, resistance has not been reported. This shows the importance that repeat applications have on eliminating sensitive wild type strains from the population.

Population Dynamics: Why a Strain Dominates in a Population

Prior to fungicide selection pressure, the original wild type sensitive strain predominated, and resistant strains which arise by mutation occurred in small numbers. The situation would remain stable until fungicide selection pressure was applied and wild type strains were reduced in numbers or eliminated by the SS fungicides. Once the wild type sensitive strains are eliminated or reduced to small percentages of the population, they have little chance to become dominant again in the population, where the fitness of the sensitive and resistant strains are equal. In other words, the situation has flip-flopped (Figure 6.3).

Population studies where fields were inoculated with various ratios of wild type sensitive strains and resistant strains of a fungus found that the strains were recovered in approximately the same percentages in which they were added. Whether the inoculum added was 50:50 wild type to resistant, 75:25 wild type to resistant, or 25:75 wild type to resistant, as long as benzimidazole fungicides were not used, the population remained essentially stable.

One thing is certain, the benzimidazole-resistant strains are equally fit and pathogenically aggressive. Once they were given a selective advantage over the wild type sensitive strain, through use of the SS fungicides, they remained the dominant strains in the population, even when the benzimidazole fungicides were no longer used.

THEORIES FOR PREVENTING OR DELAYING RESISTANCE

There are two prevalent theories on how to prevent resistance that are being promoted by uninformed people in the field today, although if challenged, proponents of these theories will concede that they may only delay resistance. It is unfortunate that either theory still exists because turf managers believe them, which has lured them into a sense of false security—of believing if they alternate SS fungicides with MS fungicides, or apply them together, that resistance will not develop. Furthermore, it has excluded them from thinking or using alternative methods of preventing or delaying resistance.

COMMONLY ACCEPTED STRATEGIES TO PREVENT OR DELAY RESISTANCE

1. Always use a combination of site-specific (systemic) fungicide with nonsite-specific (contact) fungicide.
2. Alternate a site-specific (systemic) fungicide with a nonsite-specific (contact) fungicide.

Unfortunately, neither of the theories has worked very well for a variety of reasons. For either of these theories, one has to assume the MS fungicide will eliminate any resistant strains that develop to the SS fungicides. The only problem with this assumption is that the MS fungicides, for all the years they were used, never eliminated the wild type strains or altered the makeup of the fungal population. Why should they be expected to eliminate an SS-resistant strain? MS fungicides function to suppress existing populations, not eliminate or select one strain over another. So the theory that if SS-resistant strains develop, the alternate use of an MS fungicide will eliminate that strain, has not worked. The other theory, that applying the SS and MS fungicides in combination so the MS fungicide will eliminate any SS fungicide resistant strains has not worked for the same reason.

The other problem with this theory is that for a period of time the fungal population is exposed solely to the pressure of the SS fungicide, whether they are alternated or applied together. Even if the MS fungicide is applied in combination with an SS fungicide, the MS fungicide normally lasts 10–14 days, whereas SS fungicides last 21–28 days, leaving a 7- to 14-day period of fungicide protection and population selection pressure only with the SS fungicide.

STRATEGIES TO PREVENT OR DELAY RESISTANCE TO DMI FUNGICIDES

Unfortunately, the combination theory is still being advocated with these fungicides. In addition, a philosophy of applying back-to-back applications of DMI fungicides, one at normal rate and a second at a higher rate to try to prevent resistance from developing, is also recommended. The second application at the higher rate is supposed to eliminate any low-level DMI-resistant strains that may have developed. Unfortunately, they might just be successful, and instead of having low-level DMI-resistant strains, they will undoubtedly select out high-level DMI-resistant strains! Fungi have survived for millennia and will continue to do so. Adding selection pressure to a fungal population only causes a shift away from the wild type strains to the DMI-resistant strains. This strategy will only ensure a faster elimination of DMI fungicides from your arsenal of effective fungicides for the control of diseases like dollar spot.

NEW STRATEGIES TO DELAY OR PREVENT RESISTANCE

Systemic Fungicide Reduced Rate Theory to Avoid Resistance

Sanders and her group[128] have shown that *Pythium* blight can be managed with half rates of Banol-Subdue, Aliette-Subdue, and Aliette-Banol or one-third rates of Subdue-Banol-Aliette, while preventing metalaxyl resistant strains from increasing in the population. It certainly employs the concept of low rates. There have been no reported cases of resistance to any of these fungicides in *P. aphanidermatum* where this strategy has been used. The strategy has worked very successfully so far, and since we are dealing with a facultative parasite, should continue to work in the future. Although the resistance of *P. aphanidermatum* to metalaxyl has developed slowly, it will continue to increase in future years if sound resistance strategies are not employed.

Yearly Rotation Theory to Avoid Resistance

In spite of all the concern about resistance, the main problems in turf have occurred with *S. homoeocarpa*, the cause of dollar spot, developing resistance to the various groups of fungicides. The resistance in *M. nivale* has been confined to the Pacific Northwest.[17] Since most of the

Table 6.6. Fungicide Program to Delay Development of DMI Resistant Strains

Spring-Early Summer	Mid Summer	Late Summer-Fall
DMI Fungicides	(1) Contacts (2) Contacts & Benzimidazoles (3) Benzimidazoles & Dicarboximides	(1) DMI Fungicides (2) Dicarboximides (3) Contacts

(1) indicates year 1 applications.
(2) indicates year 2 applications.
(3) indicates year 3 applications.

problems with resistance to the systemic fungicides have occurred with dollar spot, we need to develop a system to delay dollar spot resistance to the DMI fungicides. Although dollar spot can occur all season long, it is most severe in the late summer to early fall. By avoiding the heavy use of DMI fungicides during this period we should reduce the selection pressure on the population. This should allow the wild DMI-sensitive strains of *S. homoeocarpa* to remain dominant in the population.

On golf courses where DMI-reduced sensitivity strains of *S. homoeocarpa* are not present, a program like the one in Table 6.6 should help delay their appearance. This program incorporates the use of the DMI fungicides in the spring and early summer for the management of summer patch and take-all patch. The contact fungicides (chlorothalonil, anilazine), the benzimidazoles (benomyl, thiophanate methyl), and the dicarboximide fungicides (iprodione, vinclozolin) are used for the management of brown patch and anthracnose during the midsummer period. The DMI fungicides are used on a three-year rotational basis during the time of year when there is heavy dollar spot pressure. Hopefully, this will delay the dominance of DMI-resistant strains of dollar spot.

On golf courses that already have strains of *S. homoeocarpa* with reduced sensitivity to DMI fungicides, the recommendations in Table 6.6 should help delay the selection of highly resistant strains. The program incorporates the use of the DMI fungicides in the spring and early summer for summer patch and take-all patch management. It incorporates the use of benzimidazole and/or contact fungicides for the management of brown patch and anthracnose, but eliminates the use of the DMI fungicides when the incidence of dollar spot is most severe. This will hopefully put minimal pressure on the wild type dollar spot population, and allow it to remain dominant in the population.

Whether these new strategies work or not only time will tell. We should have learned by now that alternating SS fungicides and MS fungicides during the same season will not work; always applying an

MS fungicide with an SS fungicide will not work; and increasing rates and shortening intervals will not work. At least it is time to try a new strategy based on research and knowledge gained from past experiences.

SAFE HANDLING OF FUNGICIDES

Chemical Storage

Fungicides are among the safest pesticides, but they are nonetheless pesticides and should be treated accordingly (see Table 6.7 and Figures 6.4 and 6.5 for toxicity data). They should be stored in a room that is well ventilated and away from areas where people are apt to congregate. The preferable storage location would be a separate building, or at least a building that is used only to store equipment. For exact requirements, consult OSHA (Occupational Safety and Health Administration) and FEPCA (Federal Environmental Pesticide Control Act) regulations. Remember—it is your life, and inhaling those pesticide fumes daily cannot do anything but shorten it.

Proper Clothing

When handling pesticides you may need to wear latex gloves, boots, goggles, a respirator (face mask), and approved protective clothing. After application of the pesticides, the clothing should be thoroughly washed or discarded if disposable. People still insist on wearing leather boots and cloth or leather gloves; however, leather absorbs and retains pesticides. Every time your boots or gloves become wet, whether from water or sweat, the pesticides in the leather can come in contact with your skin and be absorbed by your body.

CRITERIA FOR SELECTING A FUNGICIDE

The first criterion should be how well the fungicide works (Table 6.8); that is, which is the best fungicide for the disease problem? The second criterion should be the safety of the product vis-à-vis the turfgrass. Most fungicides, properly applied, will not harm the grass, but some fungicides have greater margins of safety than others. For example, PCNB is an excellent fungicide for the control of brown patch, but application of PCNB to cool season grasses when the temperature is in the 80s ($\sim 26.6°C$) may cause yellowing and death of the

Table 6.7. Oral and Dermal Toxicity Ratings for Turfgrass Fungicides and Nematicides

	LD_{50}(mg/kg)	
	Oral	Dermal
Class 1 Highly toxic	1–50	1–200
Class 2 Moderately toxic	50–500	200–2,000
Class 3 Low order of toxicity	500–5,000	2,000–20,000
Class 4 Very low order of toxicity	over 5,000	over 20,000

Warning Statement on Label	Probable Oral Lethal Dose for 150-lb (330 kg) Person	
Class 1. Poison	Few drops to 1 tsp	.031 to 62 mL
Class 2. Danger	1 tsp to 1 oz	.62 to 29.57 mL
Class 3. Warning	1 oz to 1 pt (1 lb)	29.57 to 473.12 mL (454 g)
Class 4. Caution	1 pt to 1 qt (2 lb)	473.12 to 946.24 mL (908 g)

Generic Name	Oral LD_{50}[a]	Dermal LD_{50}[a,b]	Type[c]
Anilazine	>5,000	>5,000	F
Benomyl	>10,000	>10,000	F
Chloroneb	>11,000	>5,000	F
Chlorothalonil	>3,750	>10,000	F
Dibromochloropropan	172	1,420	N
Fenarimol	>10,000	>10,000	F
Fosetyl-Al	4,600	>2,000	F
Iprodione	>5,000	>2,000	F
Maneb	6,750–7,500	<1,000	F
Mancozeb	8,000	8,000	F
Metalaxyl	1,290–3,000	>2,000	F
Oxymal	5.4–37	2,960	N
PCNB	1,650–2,000	M	F
Propiconazole	1,310	>5,010	F
Thiram	780	S	F
Terrazole	2,000	M	F
Triadimefon	1885	>5,000	F
Phenamephos	8–9	72–84	N
Propamocarb	2,000–8,550	>3,920	F
Vapam	1,260	4,640	F,N
Vinclozolin	>5,000	>4,000	F

[a]LC_{50} = Dosage that would be fatal to one-half of those exposed.
[b]S = severe reaction, M = mild reaction
[c]F = fungicide, N = nematicide

treated turf. If another fungicide that is safer will do the job, then it should be used. The third criterion should be the applicator's safety in handling the product. The fourth criterion is the cost per unit area of applying the fungicide.

Ease of Handling

The fungicide should go into suspension or solution readily at all temperatures and should not settle out in the spray tank.

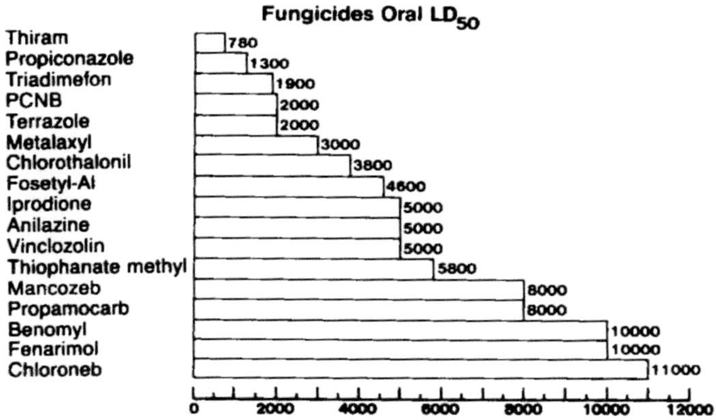

Figure 6.4. LD_{50} oral toxicity ratings (mg/kg of body weight) for turfgrass fungicides.

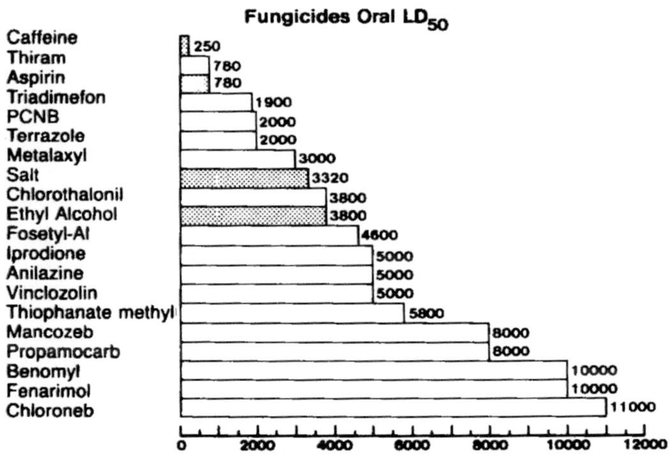

Figure 6.5. LD_{50} oral toxicity ratings (mg/kg of body weight) for turfgrass fungicides compared to commonly used ingredients.

Sprayer Calibration

Sprayer calibration is really quite simple, and yet it is one of the most difficult subjects to teach. The difficulty in teaching or learning it can only be related to the fact that mathematical calculations are involved, and the very thought of math sends instant fear through many people. The calculations are quite easy using the formulas in this section to calibrate a sprayer.

Calibrating a sprayer isn't just something that is nice to do, it is a necessity! I know of no other way to accurately apply the proper

Table 6.8. Fungicide Efficacy on Various Turfgrass Diseases

	Dollar Spot	Brown Patch	Anthracnose	Pythium Blight	Summer Patch	Necrotic Ring Spot	Take-All Patch	Melting-out	Stripe Smut	Yellow Tuft	Red Thread	Pink Patch	Powdery Mildew	Rusts	Microdochium Patch	Typhula Blight
Aliette	–	–	–	E	–	–	–	–	–	F	–	–	–	–	–	–
Banner	E	NC	E	–	E	E	E	F	E	–	E	E	E	E	E	F
Banol	–	–	–	E	–	–	–	–	–	NC	–	–	–	–	–	–
Bayleton	E	NC	E	–	E	NC	NC	NC	E	–	E	E	E	E	E	F
Bromosan	F	E	E	–	NC	NC	NC	F	NC	–	E	E	F	F	E	F
Calo Clor	–	–	–	–	–	–	–	–	–	–	–	–	–	–	E	E
Calo Gran	–	–	–	–	–	–	–	–	–	–	–	–	–	–	E	E
Chipco 26019	E	E	F	–	F	G	NC	E	NC	–	E	E	NC	NC	E	G
Clearys 3336	NC	G	E	–	G	G	G	NC	F	–	E	E	G	G	E	NC
Curalan	E	G	F	–	NC	NC	NC	E	NC	–	E	E	NC	NC	E	G
Daconil 2787	E	E	G	–	NC	NC	NC	E	NC	–	E	E	NC	NC	G	E
Duosan	NC	E	G	–	F	F	F	NC	F	–	G	G	F	F	E	NC
Dyrene	E	G	NC	–	NC	NC	NC	E	NC	–	F	F	NC	NC	F	NC
Fore, Formec	NC	E	G	–	NC	NC	NC	E	NC	–	G	G	NC	NC	E	NC
Koban	–	–	–	E	–	–	–	–	–	NC	–	–	–	–	–	–
Pace	NC	G	F	G	–	–	–	F	–	F	–	–	NC	NC	–	NC
ProStar	–	E	–	–	–	–	–	–	–	–	–	–	–	–	–	–
Rubigan	E	NC	E	–	E	E	E	NC	E	–	E	E	E	E	E	F
Subdue	–	–	–	E	–	–	–	–	–	E	–	–	–	–	–	–
Teremec sp	–	–	–	E	–	–	–	–	–	NC	–	–	–	–	NC	–
Terraclor*	NC	E*	NC	–	NC	NC	NC	E*	NC	–	NC	NC	NC	NC	G	G
Touché	E	G	NC	–	NC	NC	NC	E	NC	–	e	e	NC	NC	E	F
Turfside*	NC	E*	NC	–	NC	NC	NC	E*	NC		NC	NC	NC	NC	G	G
Twosome	E	E	G	–	G	G	G	G	F	–	E	E	NC	NC	E	G
Vorlan	E	G	NC	–	NC	NC	NC	E	NC	–	E	E	NC	NC	E	G

*For use on warm season grasses only. Is phytotoxic on cool season grasses during warm weather.
Key: E = excellent control, G = good control, F = fair control, NC = no control, – = not applicable or not known.
Note: All agricultural chemicals recommended in this chart should be applied in accordance with the directions on the manufacturer's label as registered under the Federal Insecticide, Fungicide, and Rodenticide Act. Mention of a trademark or proprietary product does not constitute a guarantee or warranty of the product by the author or the publisher and does not imply approval to the exclusion of other products that may also be suitable. Some products recommended in this chart may not be registered on the particular disease for which a rating is given. These ratings are based on experimental fungicide trials. Such products should not be used until they are fully labeled for the specific disease. The efficacy of these products is primarily based on the author's experience and efficacy of a product may be different in other areas of the world.

amount of fungicide to a given area, and that is the principle behind fungicide application. It reminds me of the time I was called to look at a disease problem that "no" fungicide was controlling. By the time I arrived, very little turf was left on the four affected greens—mainly because the superintendent had applied four potent fungicides to the affected greens, each at a concentration of 8 ounces per 1000 square feet (254 mL per 100 square meters), in a little over a week. As nearly as I could tell, the disease in question was brown patch. I asked to see the record of the spray schedule, and sure enough, the fungicides that the superintendent used should have controlled brown patch. Besides, he said, "It only occurred on four greens." I suggested that something could be wrong with his sprayer calibration, even though it didn't seem likely at the time, since only four greens were affected. At this, I noticed a strange took on his face, and I said, "You do calibrate your sprayer, don't you?" He replied, "Of course not, nobody does. You know, it's the old case for the front nine and the old case for the back nine." I said, "No, I don't know the old case for the front nine and the old case for the back nine. How can you possibly know how much you are getting on each green, or even if what you are getting on any of them is correct?" He was sure that misapplication couldn't be the problem because two of the affected greens were on the front nine and two of the affected greens were on the back nine. So I asked him in what order they were sprayed. He told me that he sent his men out to the far end of the course on the front nine, and sprayed the back nine in the same way. The last two greens on both nines were the ones with the problems, and I suggested that this might be because the sprayer wasn't calibrated, and the people spraying had to hurry over the last two greens on each nine when they discovered that they were running out of liquid in the spray tank.

Needless to say, my suggestion fell on deaf ears. The next year he doubled his fungicide budget, but didn't calibrate his sprayer. His system didn't work any better than it had before, and fortunately, for the sake of the profession, he is no longer in it.

The superintendent who followed him believed in calibration, and for the first time in 5 years grass was maintained on those four greens. The membership believed that the new superintendent could walk on water, because he kept grass on those greens all season long. It really wasn't a matter of his walking on water, just calculating how much of it was coming out of his sprayer.

Application Rate

There are four factors that affect application rate: ground speed, pressure, type or size of nozzles, and density of liquid. You need to develop a system wherein the first three remain constant and only the last one varies.

Ground Speed. Select a speed that can be used for all areas. If you are walking while treating a green, learn to walk at a comfortable, constant speed. For most people, this is their normal walking speed.

Pressure. Spray with the lowest possible pressure that gives a full pattern to ensure proper overlap between nozzles (see Figure 6.6). The trouble with people who own high-pressure sprayers is the same as the problem with people who own high-powered cars—since they have all that power they feel a need to use it. Save the higher pressures for spraying tall trees. The higher the pressure, the smaller the droplet, and the smaller the droplet, the greater the drift problem. If you aren't concerned about your neighbor (and you should be), look at it from a practical viewpoint—what drifts away isn't giving protection where it is needed.

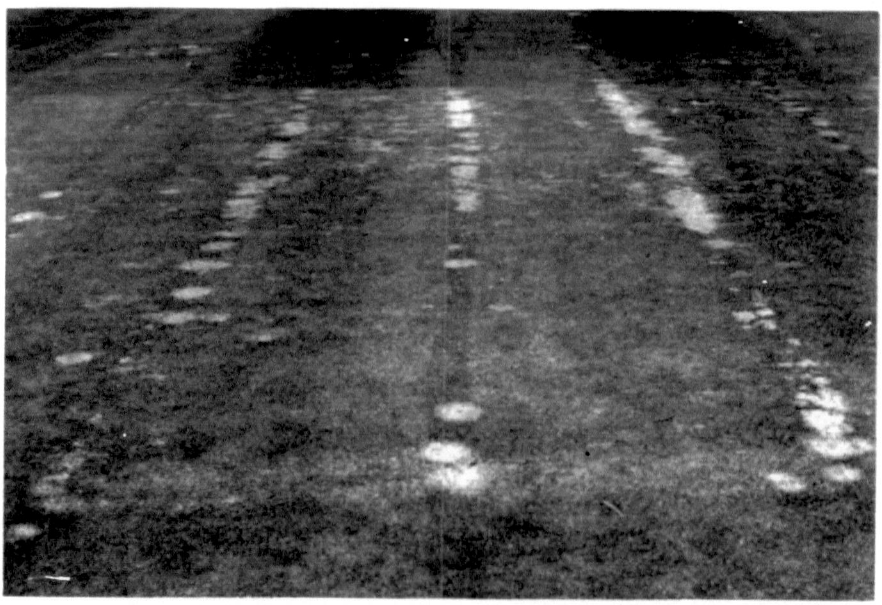

Figure 6.6. Misapplication of a granular snow mold fungicide where the spreader was not properly overlapped.

Types of Nozzles. Flat fan nozzles are still the most widely used. You should select one that you can use for all types of spraying. The nozzles should be changed yearly, because wettable powder formulations are abrasive and will enlarge or misshape the opening.

Density of Liquid. There is obviously very little that can be done to control liquid density. It will vary from one chemical to another and with concentration of fungicide. The more dense the liquid in the spray tank, the slower it will flow through the nozzles. For precise application, each chemical and rate should be checked individually.

Calibration

The purpose of calibration is to determine how much liquid your sprayer is putting out per 1000 square feet. The initial calibration usually is done with water. It is assumed that you have already selected the ground speed, the pressure, and the type of nozzles you want. Fill the tank with water, mark off 1000 square feet, and determine how much water it takes to cover the area. (This can be determined by refilling the tank with a known quantity of water.) Another method of determining sprayer output is to collect the liquid from one nozzle while covering the 1000 square feet and then multiply that volume of liquid by the number of nozzles. For either method, the process should be repeated three or four times, or until consistent output is obtained. If the rate is not satisfactory, the pressure can be raised or lowered, the ground speed changed, or the nozzles replaced with ones that have a larger or smaller opening.

Example 1. How much fungicide must be added to a 200-gallon tank to give an application rate of 4 ounces per 1000 square feet if the sprayer output is 2 gallons per 1000 square feet?

The basic output per unit area of the sprayer is the key to calibration. The output of the sprayer is 2 gallons per 1000 square feet. The amount of fungicide that must be added to the tank can then be determined by plugging figures into a simple formula.

- ounces per 1000 sq ft recommended on label
- output of sprayer (gal) per 1000 sq ft
- total ounces required in tank size or amount of water in tank

We know:
Sprayer output = 2 gal/1000 sq ft

Rate required = 4 oz/1000 sq ft
Tank size = 200 gal

We want to find out how many ounces of fungicide to add to the 200-gallon tank.

$$\frac{\text{ounces required per 1000 sq ft}}{\text{sprayer output}} = \frac{\text{fungicide needed}}{\text{tank volume}}$$

$$\frac{4 \text{ oz}}{2 \text{ gal}} = \frac{x}{200 \text{ gal}}$$

Cross multiply:

$$(2 \text{ gal})\, x = (4 \text{ oz})\,(200 \text{ gal})$$
$$2x = 800 \text{ oz}$$
$$x = 400 \text{ oz}$$

Thus, we need 400 ounces (25 pounds) of fungicide. As long as the same ground speed, pressure, and nozzles are used, the sprayer will be applying 4 ounces of fungicide per 1000 square feet. Note that we needn't know how large each green or fairway is.

Example 2. Assume that only one green of 5000 square feet needs to be treated. The sprayer output is 0.5 gallons per 1000 square feet, and the recommended rate of application for the fungicide is 2 ounces per 1000 square feet. How much water and chemical must be added to the tank?

We know:
Fungicide needed per 1000 sq ft = 2 oz
Sprayer output = 0.5 gal/1000 sq ft
Area to be sprayed = 5000 sq ft

We need to determine the amount of water to be added to the spray tank first:

$$\frac{\text{sprayer output}}{\text{unit area}} = \frac{\text{water needed}}{\text{area to be treated}}$$

$$\frac{0.5 \text{ gal}}{1000 \text{ sq ft}} = \frac{x}{5000 \text{ sq ft}}$$

$$1000x = 2500 \text{ gal}$$
$$x = 2.5 \text{ gal}$$

Thus, 2.5 gallons of water must be added to the tank to spray 5000 square feet. How much fungicide must be added to 2.5 gallons of water?

$$\frac{\text{ounces required}}{\text{sprayer output}} = \frac{\text{fungicide needed}}{\text{tank volume}}$$

$$\frac{2\ \text{oz}}{0.5\ \text{gal}} = \frac{x}{2.5\ \text{gal}}$$

$$0.5x = 5\ \text{oz}$$
$$x = 10\ \text{oz}$$

So we need 10 ounces of fungicide in 2.5 gallons of water to treat 5000 square feet, provided the same pressure and ground speed are used. (For practical purposes, 3 or 3.5 gallons of water and 12 or 14 ounces of fungicide would be added to the spray tank, because when the level in the tank becomes too low, the pump often loses its prime. Also, many sprayers cannot spray out the last gallon or so accurately.)

Example 3. There is one other type of problem that may be encountered, especially with herbicide mixtures in which different amounts of the same active ingredients (ai) are present in products sold by different companies. 2,4-D is a good example. Although this book deals only with fungicides, this information may come in handy. Since commercial 2,4-D contains various amounts of active ingredients (ai), university recommendations are often given in terms of percent ai.

How much of the 75% wettable powder (WP) formulation of benomyl is needed to get 3 ounces of ai? This is determined by dividing 100 by 75.

$$\frac{100}{75} = 1.3$$

To obtain 1 ounce of ai, 1.3 ounces of a 75% WP must be used. Since 3 ounces of ai are required, you need 3.9 ounces of benomyl, 75% WP (1.3×3 oz $= 3.9$ oz).

If the university recommendation calls for 3 ounces of active ingredient per 1000 square feet, the sprayer is calibrated to put out a gallon per 1000 square feet, and it is a 200-gallon sprayer, the amount of benomyl 75% WP that must be added to the sprayer can be determined by plugging the data into the proportions.

$$\frac{\text{ounces required}}{\text{sprayer output}} = \frac{\text{benomyl 75\% WP needed}}{\text{tank volume}}$$

$$\frac{3.9 \text{ oz}}{1 \text{ gal}} = \frac{x}{200 \text{ gal}}$$

$$x = 780 \text{ oz}$$

To apply 3 ounces of ai per 1000 square feet with a 200-gallon tank filled to capacity, you need 780 ounces of benomyl 75% WP.

The area that can be covered by one tankful of the mixture can be determined as follows:

$$\frac{\text{sprayer output}}{\text{unit area}} = \frac{\text{sprayer volume}}{\text{total area}}$$

$$\frac{1 \text{ gal}}{1000 \text{ sq ft}} = \frac{200 \text{ gal}}{x}$$

$$x = 200{,}000 \text{ sq ft}$$
(a little more than 4 acres)

CALCULATIONS TO SAVE YOU MONEY

It is difficult to teach people to buy a chemical or a fertilizer on the basis of its cost per unit area, rather than its cost per pound. I still have former students tell me what a great deal they got on a case of fungicide or a ton of fertilizer because it cost only so much a pound or a ton. When I ask them how many square feet or how many acres it will treat, too often the answer is, "What difference does it make?" Well, it makes a lot of difference. Let me illustrate with some examples.

Example 4. Fungicide A costs $2.00 per pound; fungicide B costs $10.00 per pound, and both will manage dollar spot. Fungicide A, at $2.00 per pound, appears to be the better buy. But is it?

The cost that counts is the cost per 1000 square feet, not the cost per pound. Assume fungicide A is applied at a rate of 8 ounces per 1000 square feet and fungicide B at a rate of 1 ounce per 1000 square feet. Since there are 16 ounces in a pound, the cost per ounce of fungicide A is $2.00/16, or 12.5 cents, and the cost per ounce of fungicide B is $10.00/16, or 62.5 cents. At a rate of 8 ounces per 1000 square feet, fungicide A would cost 12.4 cents × 8, or $1.00 per 1000 square feet.

Fungicide B, applied at a rate of 1 ounce per 1000 square feet, would cost 62.5 cents per 1000 square feet. Thus, fungicide B turns out to be the better buy.

Remember, it is the cost per unit area that you consider when buying fertilizer and chemicals, not the cost per pound! By following the steps in Example 4, you can figure out cost per 1000 square feet. If you don't want to do the figuring yourself, ask your salesman to do it for you. (If he can't, you need a new salesman.)

Consider one more example; this time with fertilizer. It is with fertilizers and herbicides that people most often make the mistake of buying on the basis of cost per pound instead of cost per 1000 square feet.

Example 5. Fertilizer A is a 15–0-0 mixture that costs $150.00 per ton, and fertilizer B is a 35–0-0 mixture that costs $300.00 per ton. You want to apply 1 pound of actual nitrogen per 1000 square feet. Which fertilizer is the better buy?

The numbers 15 and 35 on the labels give the percentage of nitrogen per pound. To get 100% (1 pound) actual nitrogen you must apply 100/15, or 6.67 pounds of fertilizer A. With fertilizer B you will need 100/35, or 2.86 pounds.

Since there are 2000 pounds in a ton, the cost per pound of fertilizer A is $150.00/2000, or 7.5 cents, and the cost per pound of fertilizer B is $300.00/2000, or 15 cents. The cost per 1000 square feet of fertilizer A is 6.67×7.5 cents, or 50 cents; the cost per 1000 square feet of fertilizer B is 2.86×15 cents, or 43 cents. Thus, fertilizer B is cheaper to use, even though it costs more per ton.

When buying fertilizer you must be aware that there are different forms of nitrogen and take that into account. The point I'm making here, however, is that you should buy fertilizer according to the cost of actual nitrogen per unit area, not the cost per ton.

Calibration (Metrics)

The purpose of calibration is to determine how much liquid your sprayer is putting out per 100 square meters. The initial calibration usually is done with water. It is assumed that you have already selected the ground speed, the pressure, and the type of nozzles you want. Fill the tank with water, mark off 100 square meters, and determine how much water it takes to cover the area. (This can be determined by refilling the tank with a known quantity of water.) Another method of determining sprayer output is to collect the liquid from one nozzle while covering the 100 square meters and then multiply that volume of liquid by the number of nozzles. For either method, the process should be repeated three or four times, or until consistent output is obtained. If the rate is not satisfactory, the pressure can be raised or lowered, the ground speed changed, or the nozzles replaced with ones that have a larger or smaller opening.

Example 1. How much fungicide must be added to a 750-liter tank to give an application rate of 130 milliliters per 100 square meters if the sprayer output is 8 liters per 100 square meters?

The basic output per unit area of the sprayer is the key to calibration. The output of the sprayer is 8 liters per 100 square meter. The amount of fungicide that must be added to the tank can then be determined by plugging figures into a simple formula.

- milliliters per 100 square meters recommended on label
- output of sprayer (liters) per 100 square meters
- total milliliters required in tank size or amount of water in tank

We know:
Sprayer output = 8L/100 sq meters
Rate required = 130mL/100 sq meters
Tank size = 750L

We want to find out how many milliliters of fungicide to add to the 750-liter tank.

$$\frac{\text{milliliters required per 100 sq meters}}{\text{sprayer output}} = \frac{\text{fungicide needed}}{\text{tank volume}}$$

$$\frac{130 \text{ mL}}{8\text{L}} = \frac{x}{750\text{L}}$$

Cross multiply:

$$\begin{aligned}(8\text{L})(x) &= (130\text{mL})(750\text{L}) \\ 8x &= 97{,}500 \text{ mL} \\ x &= 12{,}188 \text{ mL}\end{aligned}$$

Thus, we need 12,188 milliliters (12.19 liters) of fungicide. As long as the same ground speed, pressure, and nozzles are used, the sprayer will be applying 130 mL of fungicide per 100 square meters. Note that we needn't know how large each green or fairway is.

Example 2. Assume that only one green of 500 square meters needs to be treated. The sprayer output is 2 liters per 100 square meters, and the recommended rate of application for the fungicide is 60 milliliters per 100 square meters. How much water and chemical must be added to the tank?

We know:
Fungicide needed = 60 mL/100 sq meters
Sprayer output = 2L/100 sq meters
Area to be sprayed = 500 sq meters

We need to determine the amount of water to be added to the spray tank first:

$$\frac{\text{sprayer output}}{\text{unit area}} = \frac{\text{water needed}}{\text{area to be treated}}$$

$$\frac{2\ \text{L}}{100\ \text{sq m}} = \frac{x}{500\ \text{sq m}}$$

$$100x = 1000\ \text{L}$$
$$x = 10\ \text{L}$$

Thus, 10 liters of water must be added to the tank to spray 500 square meters. How much fungicide must be added to 10 liters of water?

$$\frac{\text{milliliters required}}{\text{sprayer output}} = \frac{\text{fungicide needed}}{\text{tank volume}}$$

$$\frac{60\ \text{mL}}{2\ \text{L}} = \frac{x}{10\ \text{L}}$$

$$2x = 600\ \text{mL}$$
$$x = 300\ \text{mL}$$

So we need 300 milliliters of fungicide in 10 liters of water to treat 500 square meters, provided the same pressure and ground speed are used. (For practical purposes, 15 liters of water and 450 milliliters of fungicide would be added to the spray tank, because when the fluid level in the spray tank becomes too low, the pump often loses its prime. Also, many sprayers cannot spray out the last couple of liters or so accurately.)

Example 3. There is one other type of problem that may be encountered, especially with herbicide mixtures in which different amounts of the same active ingredients (ai) are present in products sold by different companies. 2,4-D is a good example. Although this book deals only with fungicides, this information may come in handy. Since commercial 2,4-D contains various amounts of active ingredients (ai), university recommendations are often given in terms of percent ai.

How much of the 75% wettable powder (WP) formulation of benomyl is needed to get 90 milliliters of ai? This is determined by dividing 100 by 75.

$$\frac{100}{75} = 1.3$$

To obtain 30 milliliters of ai, 39 milliliters of a 75% WP must be used. Since 90 milliliters of ai are required, you need 117 milliliters of benomyl, 75% WP (1.3 × 90 mL = 117 mL.)

If the university recommendation calls for 90 milliliters of active ingredient per 100 square meters, the sprayer is calibrated to put out 5 liters per 100 square meters, and it is a 750-liter sprayer, the amount of benomyl 75% WP that must be added to the sprayer can be determined by plugging the data into the proportions.

$$\frac{\text{milliliters required}}{\text{sprayer output}} = \frac{\text{benomyl 75\% WP needed}}{\text{tank volume}}$$

$$\frac{117 \text{ mL}}{5 \text{ L}} = \frac{x}{750 \text{ L}}$$

$$x = 17{,}550 \text{ mL}$$

To apply 90 milliliters of ai per 100 square meters with a 750-liter tank filled to capacity, you need 17,550 milliliters of benomyl 75% WP.

The area that can be covered by one tankful of the mixture can be determined as follows:

$$\frac{\text{sprayer output}}{\text{unit area}} = \frac{\text{sprayer volume}}{\text{total area}}$$

$$\frac{5 \text{ L}}{100 \text{ sq m}} = \frac{750 \text{ L}}{x}$$

$$x = 15{,}000 \text{ sq m}$$

CALCULATIONS TO SAVE YOU MONEY

It is difficult to teach people to buy a chemical or a fertilizer on the basis of its cost per unit area rather than its cost per kilogram. I still have former students tell me what a great deal they got on a case of fungicide or a ton of fertilizer because it cost only so much a kilogram or a ton. When I ask them how many square meters or how many hectares it will treat, too often the answer is, "What difference does it make?" Well, it makes a lot of difference. Let me illustrate with some examples.

Example 4. Fungicide A costs \$2.00 per kilogram; fungicide B costs \$10.00 per kilogram, and both will manage dollar spot. Fungicide A, at \$2.00 per kilogram, appears to be the better buy. But is it?

The cost that counts is the cost per 100 square meters, not the cost per kilogram. Assume fungicide A is applied at a rate of 250 grams per 100 square meters and fungicide B at a rate of 30 grams per 100 square meters. The cost per kilogram of fungicide A is \$2.00, and the cost per kilogram of fungicide B is \$10.00. At a rate of 250 grams per 100 square meters, fungicide A would cost 2 × 0.250, or \$.50 per 100 square meters. Fungicide B, applied at a rate of 30 grams per 100 square meters, would cost 10 × .030, or \$.30 per 100 square meters. Thus, fungicide B turns out to be the better buy.

Remember, it is the cost per unit area that you consider when buying fertilizer and chemicals, not the cost per kilogram! By following the steps in Example 4, you can figure out cost per 100 square meters. If you don't want to do the figuring yourself, ask your salesman to do it for you. (If he can't, you need a new salesman.)

Consider one more example; this time with fertilizer. It is with fertilizers and herbicides that people most often make the mistake of buying on the basis of cost per kilogram instead of cost per 100 square meters.

Example 5. Fertilizer A is a 15–0-0 mixture that costs \$150.00 per ton, and fertilizer B is a 34–0-0 mixture that costs \$300.00 per ton. You want to apply 0.5 kilograms of actual nitrogen per 100 square meters. Which fertilizer is the better buy?

The numbers 15 and 35 on the labels give the percentage of nitrogen per kilogram. To get 100% (1 kilogram) actual nitrogen you must apply 100/15, or 6.67 kilograms of fertilizer A. With fertilizer B you will need 100/35, or 2.86 kilograms.

Since there are 1000 kilograms in a ton, the cost per kilogram of fertilizer A is \$150.00/1000, or 15 cents, and the cost per kilogram of fertilizer B is \$300.00/1000, or 30 cents. The cost per 100 square meters of fertilizer A is 6.67 × 15 cents, or \$1.00; the cost per 100 square meters of fertilizer B is 2.86 × 30 cents, or 86 cents. Thus, fertilizer B is cheaper to use, even though it costs more per ton.

When buying fertilizer you must be aware that there are different forms of nitrogen and take that into account. The point I'm making here, however, is that you should buy fertilizer according to the cost of actual nitrogen per unit area, not the cost per ton.

CHAPTER 7

Cultural Aspects of Turfgrass Disease Management

Cultural management is, of course, only one aspect of a comprehensive program for managing turfgrass disease. The other major components of a disease management program, resistant cultivars and chemical management, will be discussed in Chapters 11 and 12, which describe disease management programs. This chapter deals with the effects of soil fertility, soil pH, watering, and mowing on turfgrass disease.

SOIL FERTILITY

Nitrogen

Adding nitrogen to the soil makes some turfgrass diseases worse, but reduces the severity of others (see, for example, Figure 7.1.) Some diseases in each category are listed in Table 7.1.

Once you know the effect of nitrogen on disease development, timing the nitrogen applications becomes the next important consideration. Nitrogen applied in the spring is not going to solve a stem rust problem

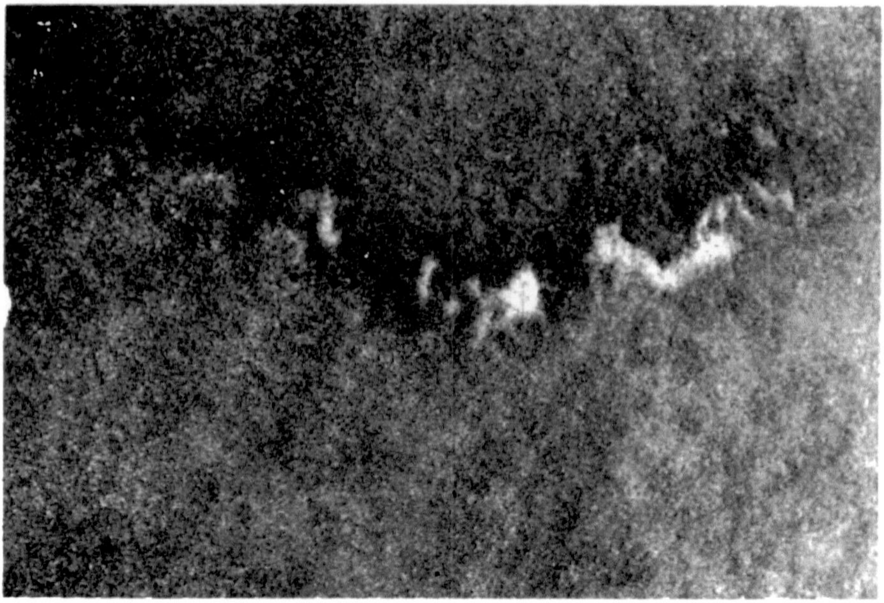

Figure 7.1. ***Microdochium*** **patch occurring in the fairy ring zone of stimulation where excess nitrogen has increased the susceptibility of the turf.**

Table 7.1. Effects of Nitrogen Application on Selected Turfgrass Diseases

Severity Increased by Nitrogen	Severity Decreased by Nitrogen
Pythium blight	Dollar spot
	Rust
Brown patch	Red thread
Gray leaf spot	Pink patch
Stripe smut	Anthracnose
Microdochium patch	Necrotic ring spot
Typhula blight	Summer patch
	Melting-out
	Leaf spot

in the fall. Timing nitrogen applications to make a particular disease less severe is not as simple as it may seem. In the course of one growing season a single turfgrass species may be subject to a variety of diseases, all of which must be taken into account. It would be ideal if each turfgrass species were susceptible to only one disease, for then it would be easy to plan nitrogen applications. Unfortunately, as things are, a program of nitrogen fertilization that alleviates one disease may worsen another.

You must balance disease management with the nutrient needs of the turfgrass plant. Grass plants need nitrogen, and so you cannot simply

Table 7.2. Recommended Annual Phosphate (P_2O_5) Applications (Bray P_1-Extractable)

Soil Test (lb P/acre) (kg/hectare)	General Turf (lb/1000 sq ft) (kg/100 sq m)	General Turf (lb/acre) (kg/hectare)	High Maintenance Turf[a] (lb/1000 sq ft) (kg/100 sq m)
Less than 15 (very low) (17)	3 (1.5)	130 (145)	4 (2)
16–25 (low) (18–28)	2 (1.0)	85 (95)	3 (1.5)
26–40 (medium) (29–45)	1 (0.5)	45 (50)	2 (1)
41–70 (high) (46–79)	0 (0)	0 (0)	1 (0.5)
More than 70 (very high) (>79)	0 (0)	0 (0)	0 (0)

[a]Includes greens and tees.

eliminate all nitrogen applications in the interests of disease management. You should apply nitrogen at the time of year when it will assist disease management the most.

Phosphorus and Potassium

After nitrogen, the two most important elements for growing plants are phosphorus and potassium. The role of these elements in controlling diseases has not been determined, although Madison[99] explains how the interaction of nitrogen, phosphorus, and potassium affects specific diseases. If most turfgrasses were subject to only one serious disease, it might be worthwhile to maintain relative concentrations of nitrogen, phosphorus, and potassium designed to mitigate that disease. But, unfortunately, most turfgrasses species have many diseases during a single growing season, and while it is fairly easy to regulate the nitrogen level, it is difficult to alter the amounts of phosphorus and potassium available to the plant. Phosphorus and potassium tend to be insoluble and thus are not readily leached. It is easy to raise the level of these nutrients in the soil but hard to bring it back down quickly. It would be very difficult to implement any program that required raising and lowering the levels of phosphorus and potassium throughout the season. The simplest solution is to maintain adequate levels of phosphorus and potassium and vary the amount of nitrogen.

What are adequate levels of phosphorus and potassium? Guidelines are given in Tables 7.2 and 7.3, which are based on recommendations from the Michigan State University Soil Testing Lab. Other labs may use different methods of testing soil.

Sulfur

Sulfur is known to be a fungicide. It may act to reduce disease directly as a fungicide or indirectly as a plant nutrient. Goss and

Table 7.3. Recommended Annual Potash (K_2O) Applications (Neutral 1 N Ammonium Acetate-Extractable)

Soil Test (lb P/acre) (kg/hectare)	General Turf (lb/1000 sq ft) (kg/100 sq m)	General Turf (lb/acre) (kg/hectare)	High Maintenance Turf[a] (lb/1000 sq ft) (kg/100 sq m)
Less than 15 (very low) (17)	4 (2)	170 (190)	5 (2.5)
15–101 (low) (17-113)	3 (1.5)	130 (145)	4 (2)
101–175 (medium) (113–197)	2 (1)	85 (95)	3 (1.5)
175–250 (high) (197–280)	1 (0.5)	45 (50)	2 (1)
More than 250 (very high) (>280)	0 (0)	0 (0)	1 (0.5)

[a]Greens, tees, and other high maintenance turfs on sandy soils with high irrigation rates.

Gould[63] demonstrated control of *Microdochium* patch with sulfur, and Dernoeden showed take-all patch was reduced by sulfur.[32] While all this may be well and good, there are other fungicides and cultural practices that can be used to manage these diseases without the potential problems that sulfur poses. Sulfur has been shown to be the primary cause of the black layer. Managing *Microdochium* patch and take-all patch with sulfur is a little like curing the common cold with chemotherapy: it might do the job but God help the patient. These two diseases can be managed using less drastic means, but it is very difficult to manage a black layer once it occurs. (See Black Layer section, Chapter 5).

Iron

Iron is a minor element that is applied when turf becomes chlorotic because of an iron deficiency. This deficiency may occur when low levels of iron are present in the soil, or when adequate levels of iron are present in the soil, but unavailable due to the high soil pH. When soluble iron is applied to foliage, it is a matter of only half an hour or so before the turf looks darker and greener. Iron sulfate is generally used, although other forms of soluble iron are available. If you desire a darker green turf, adding iron is a much safer way to get it than applying nitrogen. Adding excessive amounts of nitrogen will only encourage development of certain diseases, and you may be left with brown turf or bare ground instead of the lovely dark green grass you envisioned.

Iron sulfate may be used to mask the symptoms of yellow tuft disease. While this won't control the problem, it will make turf infected with yellow tuft look better.

SOIL pH

Most of the literature tells you that the soil pH should be maintained at the optimum level for turfgrass growth (between 6 and 7).[4,149] This is really part of folklore, and is based on research done on wheat and corn. Having a pH between 6 and 7 is important so the wheat and corn plant can have the phosphorus in the soil available at heading time because it is difficult to run a fertilizer down the field when the corn is six feet high. On the other hand, you can run a fertilizer spreader over your turf every day if need be. It is much safer to add the fertilizer as needed than to try to lower the pH with sulfur and wind up with a black layer. In most instances it is also impossible to lower the pH of soils above 7.5 with sulfur because of the high buffering capacity of the soil (see Black Layer section, Chapter 5).

Although some diseases do respond to changes in soil pH, it is not practical to combat turfgrass diseases by adjusting pH. For example, it is impossible to change the soil pH from 5 to 7 to fight one disease, and then change it back again when a different disease comes along. Since the effect of pH on turfgrass diseases is usually related to the levels of nitrogen, phosphorus, and potassium, it is much simpler to live with the pH you have, adjust the levels of phosphorus and potassium in the soil, and add nitrogen as needed.

IRRIGATION

Proper irrigation can help minimize turfgrass diseases. The best time to irrigate is in the afternoon, lightly, and daily. It not only supplies the plant with water, but helps cool it off so it can better make it through the heat stress of midday. The worst time to irrigate turf is early to late evening. This wets the turfgrass plant and debris (mat and thatch) and allows foliar pathogens to germinate, grow, and infect all night, since normally very little drying takes place before sunrise. Watering early in the evening also cools off the plants and promotes the formation of guttation water, which is rich in nutrient and encourages even more disease development.

If irrigating during the day is impractical, the second best time to irrigate is just before sunrise. The water dilutes the nutrient-rich guttation water and, by breaking up the droplets, allows quicker drying after the sun rises.

DRAINAGE

Good drainage is just as important as proper irrigation. In the transition zone and the warm-season grass areas, *Pythium* blight requires chemical management, but in the northern region of the cool-season grass belt, *Pythium* blight indicates a drainage problem. If you correct the drainage problem you will also clear up most of the *Pythium* blight. Other diseases that are made worse by poor drainage are brown patch, and *Typhula* blight.

Air drainage is important, too. Diseases like *Microdochium* patch, *Pythium* blight, brown patch, powdery mildew, and gray leaf spot are more severe where air drainage is poor. Removing or pruning some trees, especially the lower limbs, will increase the air circulation and make these diseases less severe.

MOWING

Height of Cut

Grass should not be mowed shorter than its minimum competitive mowing height (see Table 7.4).[4,149] For some species, like creeping bentgrass, this can be as short as 1/8 inch (0.3 cm), although 3/16 inch (0.5 cm) is more practical, especially if you want to prevent the creeping bentgrass from becoming annual bluegrass, and 1/2 inch (1.3 cm) would be preferable (but is not possible on greens). Other species—Kentucky bluegrass, for example—have a minimum competitive height of 1 inch (2.5 cm), and certain cultivars must be cut even higher. It has been suggested that the "new" Kentucky bluegrasses could be cut shorter and still compete with other grasses such as annual bluegrass. Maybe they could, if they didn't have to sustain traffic, but traffic can't be eliminated from a golf course fairway. There is no reason to mow home

Table 7.4. Mowing Heights of Some Turfgrass Species[4,149]

Species	Minimum Height (in) (cm)	Preferred Height (in) (cm)
Creeping bentgrass	1/8 (0.3)	1/4–1 (0.6–2.5)
Kentucky bluegrass	1 (2.5)	2–3 (5–7.6)
Fine-leaf fescue	1/2 (1.3)	2–3 (5–7.6)
Bermudagrass	1/8 (0.3)	1/2–3 (1.3–7.6)
St. Augustinegrass	3/4 (1.9)	2–3 (5–7.6)
Zoysiagrass	3/4 (1.9)	2–3 (5–7.6)
Annual bluegrass	1/8 (0.3)	1/8–1 (0.6–2.5)

lawns so short. Turfgrass plants mowed shorter than their optimal height of cut are, in general, more susceptible to diseases.

Seasonal variation in mowing heights can be beneficial. A lower cutting height during cool weather retards thatch development, whereas higher mowing heights during warm weather may lower the temperature around the crowns and help the grass to survive the stress period.

Frequency

How often you mow depends on how much the turf is used. In any case, you must make sure that no more than one-third of the shoot growth is removed during a single mowing. A golf green needs daily mowing (at least 6 days a week), and fairways and tees require mowing daily to once a week (three times a week is common). For areas of general use such as parks, home lawns, and athletic fields, mowing once a week is usually sufficient.

Mowing makes wounds through which pathogenic fungi can enter the plant and infect it. The more you mow, the more fresh wounds the grass will have. A dull mower inflicts more and bigger wounds than a sharp mower. Wounds made by a sharp mower are cleaner and heal faster than the tearing and shredding caused by a dull mower.

Clippings

It is not clear what effect clippings have on disease development. There is little evidence to support the theory that clippings left on the ground increase inoculum levels and thereby encourage disease. In fact, clippings seem to have little direct effect on disease development, and they do not contribute to thatch. With some diseases, like melting-out, large amounts of inoculum may be present in the crown area and root zone; however, there is little evidence that the disease is worse where clippings are left than where they are removed. Even when clippings are removed, there appears to be plenty of inoculum for an epidemic. If you want to prevent a melting-out problem, plant a resistant grass variety and leave the clippings debate to the plant pathologists.

Golf course greens have more diseases than other turfs although the clippings are removed. For diseases with airborne inoculum, like the rusts, smuts, powdery mildews, and leaf spots in general, removing the clippings won't affect the inoculum level.

Leaving the clippings does, however, affect the total nitrogen and potassium available to the plant. Turfs from which the clippings are removed require more added fertilizer than turfs on which clippings are left.

Leaving the clippings on a higher cut home lawn does not appear to cause any detrimental effects to the turf, assuming the turf is mowed frequently enough so that no more than $^1/_3$ of the plant is removed at any one time. In fact, with the looming crisis of running out of landfill space, homeowners may find it impossible to dispose of their clippings and may have to resort to mulching mowers.

It is a different story on golf courses, where removing clippings definitely improves the quality of creeping bentgrass-annual bluegrass greens and fairways. On fairways, removing the clippings during the warm weather of summer has definitely improved the quality of the fairways. They remain much denser than fairways where clippings are not removed. No one seems to know exactly why, but it appears to be related to the rapid breakdown of the clippings in the warm weather, releasing toxic substances that thin the turf.

There has also been a noticeable increase in the creeping bentgrass content on fairways where clippings are removed. It appears that where the creeping bentgrass is not thinned by the toxins in the clippings, it gains a competitive advantage over the annual bluegrass during the warm weather.

Collecting Grass Clippings to Prevent Thatch

This theory should rank with such fairy tales as Peter Pan, Goldilocks and the Three Bears, and Cinderella. God alone probably knows who started the tale, but few have spread as fast and as widely. Good experimental data have disproved the theory that grass clippings are a contributor to thatch development, but common sense should have disproved the theory long ago. Few areas have the thatch problems that a golf course green has, and yet clippings are removed there all the time! Collecting the clippings just removes valuable nitrogen and potassium. An additional 1 to 2 pounds of nitrogen per 1000 square feet (.5 to 1 kg per 100 square meters), as well as supplemental potassium, will have to be added each season where the clippings are removed. Thatch is composed primarily of rhizomes, stolons, and roots, which you don't mow. What you do mow is the leaf blades, which are composed of materials, cellulose and hemi-cellulose, that are broken down readily by microorganisms. The other plant parts have a high lignin content, are therefore not readily broken down by microorganisms, and consequently contribute to the development of thatch.

Does Thatch Cause Disease?

Many people say that thatch causes disease, but there is little evidence that it actually does so. It is suggested that thatch harbors the pathogens that cause turf diseases. However, many diseases are airborne, like the rusts and powdery mildew, and the rest do just as well in a soil environment as in a thatch environment. The one problem with turfs that are maintained in thatch is that they are more susceptible to drought than are turfs that are maintained in soils. Consequently, diseases like necrotic ring spot, summer patch, and stripe smut, which develop symptoms under conditions of drought stress, will be more severe. However, this is quite different from the theory that thatch is the place that harbors all those nasty pathogens, as proposed by many turfgrass experts. The suggestion that if we somehow eliminate the thatch, all our disease problems would go away, just isn't true. Believe me, that's not going to happen.

Removal of the clippings for a crown- or root-rot disease, like necrotic ring spot and summer patch, has little effect on inoculum levels. There are enough airborne spores of the fungi that cause melting-out, rusts, powdery mildew, and smuts to negate what little could be done by removing clippings. The perfect example again is golf course greens, where clippings are always removed, yet there are as many disease problems on golf course greens as there are anywhere else.

Should you dethatch your lawn? The answer to that question depends on how much weight you have gained during the winter. If you gained a lot of weight, the exercise of running the power rake might be good for you. On the other hand, if you are too far out of shape, it might kill you! But it will do nothing for your thatch problem. You will remove the winter-killed grass, which would break down anyway, but you will not remove that thatch or correct a thatch problem. Where severe thatch exists, the lawn should be cored (aerated) and the soil in the cores returned to help facilitate thatch breakdown. It may be necessary to repeat this operation several times where severe thatch problems exist.

Disease Management by Mowing

Rust, red thread, melting-out, and leaf spot can be managed by mowing. Recommendations for managing these diseases call for increasing the nitrogen level. However, if you add nitrogen and do not mow, you will actually make the disease worse, not better. This has been demonstrated many times in the laboratory. In the field, where mowing occurs at least once a week, infected foliage is mowed off before the above-mentioned fungus has a chance to complete the disease cycle. Since these pathogens have a 10–14 day cycle from infection to sporulation, mowing once a week keeps it from becoming a serious problem.

CHAPTER 8

Using Disease Resistance in Turfgrass Management

Diversity is the rule of nature. Rarely, whether in grassland or forest, is an extensive monostand found. Most of the single-species plant communities that exist today have been purposely developed to satisfy agricultural or aesthetic needs. Monocropping has been successful with annual crop plants, especially since the development of selective herbicides, but it has proved to be more difficult to establish and maintain single stands of perennial turfgrasses. Mixtures of turfgrass species and blends of turfgrass cultivars are now being used in efforts to imitate the patterns of nature. In this chapter, the reasons for blending and mixing will be examined, and the two forms of resistance, horizontal and vertical, explained.

MIXTURES AND BLENDS

A **mixture** consists of two or more turfgrass species planted together; a **blend** is two or more cultivars of the same species planted together.[4,149] The purpose of using a mixture or a blend is to create a diversified plant community which will have greater resistance to stress and disease, and thus greater longevity, than a stand composed of a single species or cultivar. Simply mixing any two species or cultivars, however, will not necessarily improve quality or increase longevity. Appropriate species or cultivars must be selected. Mixing a fine-leaf fescue with a common Kentucky bluegrass (*Poa pratensis*) in a home lawn, for example, results in a plant population capable of producing acceptable turf under a range of environmental stresses. Thus, Kentucky bluegrass will usually predominate in areas that receive full sun, and the fine-leaf fescues will thrive in shaded or droughty areas. Although theoretically this sounds great, the mixture does not always produce a uniform turf. Sometimes the fine-leaf fescue tends to persist in the areas of full sunlight, often as individual clumps or patches which detract from the lawn's appearance.

The principle of blending cultivars to achieve disease-resistant turf is applicable to all turfgrasses. In the cool-season regions, Kentucky bluegrass blends are the most common, and therefore most of the following information deals with that species, but the principles of blending apply to the other grasses as well.

Blends vs. Single Cultivars

Until the mid-1960s, most universities and commercial seed companies recommended planting single cultivars, particularly if high quality was desired. Then, in the mid-1960s, researchers began to blend Kentucky bluegrass cultivars, with an apparent improvement in turfgrass quality. Reevaluation suggests, however, that the quality of blended turf may not be better at all. (Data are available in turfgrass conference reports and turfgrass field reports from several universities.) The sort of misinterpretation that has occurred is shown by the following hypothetical example evaluating resistance to melting-out in Kentucky bluegrass. Experimental results like those in Table 8.1 have been reported as demonstrating that a blend of Merion and common Kentucky bluegrass is superior to common Kentucky bluegrass. The quality rating was increased from 8 for common Kentucky bluegrass alone to 4 for the blend, and the percentage of infection was reduced from 70% for

Table 8.1. Hypothetical Melting-out Ratings

	Percentage of Turf Infected	Quality Rating (1 = best, 9 = poorest)
Merion	10	2
Common	70	8
Merion-Common blend	40	4

Figure 8.1. Left-hand side of sod field is planted to the single melting-out resistant Kentucky bluegrass cultivar 'Merion,' whereas the right-hand side was planted to a blend of melting-out susceptible Kentucky bluegrass cultivars 'Prato,' 'Delta,' and 'Park,' showing the importance of blending the right cultivars.

common Kentucky bluegrass alone to 40% for the blend. Overlooked, however, was the fact that Merion alone had a quality rating of 2, with only 10% infection. The blend is indeed an improvement over common Kentucky bluegrass, but it is inferior to Merion alone. (For more information on the averaging effect of blending, see Vargas and Turgeon.[161])

Blending will not cure all turfgrass ills. A blend of two improperly selected cultivars may produce a less desirable turf than a single superior cultivar would (see, for example, Figure 8.1).

"Super" Blends

The theory that a blend can be made that will suit all situations is untenable. Blends must be developed for specific uses, such as high-maintenance turf, low-maintenance turf, or shaded turf. Attempting to develop a blend that meets all those needs will result in fair to poor turf for all areas, because the cultivars that are not adapted to the environment and cultural regime of an area will deteriorate, lowering the turf quality. The preferred practice is to seed or sod with a blend developed for a particular environment and cultural regime. Small exceptions, such as shaded portions of a home lawn, can be seeded or sodded with an appropriate single cultivar instead of with the blend.

Should Blends Include a Common Type?

There is a theory that blends of Kentucky bluegrass should include a common type along with the improved types.* Two reasons for including a common type are (1) it keeps the cost of seed down, and (2) if the seed or sod receives low maintenance, the common type will predominate. Common Kentucky bluegrasses, however, are susceptible to melting-out; the improved types are resistant. (The terms "common" and "improved" are open to question.) There is, thus, an obvious danger in adding the susceptible grasses to a blend.

Vargas and Turgeon[161] found that blending common types like Kenblue and Park with the resistant cultivars Merion, Nugget, and Pennstar produced turf of lower quality than stands of single resistant cultivars. After 4 years, the blends Pennstar-Kenblue, Pennstar-Park, Nugget-Kenblue, Nugget-Park, and Merion-Park were still of intermediate quality during the time of year when melting-out was active, indicating that the common types had not been eliminated from the blends. (This probably happened because the large amounts of inoculum produced on the susceptible cultivars broke down the resistance of the improved cultivars, preventing them from gaining a competitive advantage.) The hypothesis that melting-out would quickly eliminate the common type from such blends was thus disproved. Only in the Merion-Kenblue blend was there an indication that the common type, in this case Kenblue, was being eliminated from the blend. This was observed in the fourth season. Some might conclude that in this case the combination was justified, yet for 4 years the turf has been of lower quality than Merion itself, and this condition undoubtedly will continue for at least another 2 years until all of the Kenblue has been eliminated. The result of the Merion-Kenblue blend is a lawn of intermediate quality for 6 years. The normal life span of a first-class turf given no renovation (to remove perennial weedy grasses) is 5 to 10 years. Thus 6 years have been wasted. Moreover, the weakened condition of the common component during the spring, when melting-out occurs, allows perennial weedy grasses to invade. When the Kenblue is finally eliminated from the blend, what will be the result? A monostand of Merion! If that was the goal, what was the purpose of the blend?

* The term "common" applied to Kentucky bluegrass denotes susceptibility to melting-out; the term "improved" denotes resistance to the disease.

Even in a blend of two resistant cultivars and one susceptible cultivar, if the susceptible cultivar is eliminated after a period of time, the genetic diversity is reduced by one-third, and poorer quality has been tolerated while this was occurring. A blend of three disease-resistant cultivars would have given a lawn of superior quality, and the genetic diversity contributed by all three cultivars would have been present for the life of the turf.

HORIZONTAL AND VERTICAL RESISTANCE

Plants have two types of resistance to disease: specific and generalized.[152] Specific resistance (also called vertical resistance) occurs when plant *X* is bred or selected for resistance to a specific race of a pathogen; for example, race 1 of pathogen *a*. Generalized, or horizontal, resistance occurs when plant *X* is bred or selected for resistance to all existing races of a particular pathogen or without regard to whether races of the pathogen exist. Specific resistance in its highest form is immunity, or complete freedom from infection. Generalized resistance is usually less than immunity, but results in fewer infections, smaller lesions, longer intervals between infection and sporulation, and fewer spores. These results are brought about by a reduction in the amount of inoculum.

Specific resistance is most effective against pathogens lacking a sexual stage and having only one life cycle a year. A pathogen that has a sexual cycle, or that has repeating asexual cycles throughout the growing season, is more likely to develop a race able to overcome specific resistance. Generalized resistance, on the other hand, is effective against pathogens whether or not they have a sexual cycle or a repeating asexual cycle. The possible development of new races is not important, since generalized resistance is resistant to all races of a pathogen.

The following discussion treats resistance in the context of three important diseases of Kentucky bluegrass: melting-out, stripe smut, and necrotic ring spot.

Melting-Out

The successful development, through hybridization and selection, of cultivars of Kentucky bluegrass resistant to melting-out is a classic example of generalized resistance. Merion Kentucky bluegrass was the first melting-out-resistant cultivar to be discovered and developed. The fact that its resistance has been sustained for over 30 years shows how effective horizontal resistance is. A number of Kentucky bluegrass cultivars developed since then also have excellent resistance to melting-out.

If two cultivars, one susceptible and one resistant to melting-out, are blended, you might expect that the susceptible cultivar would become infected and the resistant one would not. But remember that horizontal resistance means a lowered inoculum level (see Figure 8.2). A resistant cultivar such as Merion tends to keep the inoculum level low during the 4 to 5 weeks that melting-out is active, but the presence in the blend of a susceptible common type like Park nullifies this effect by bombarding

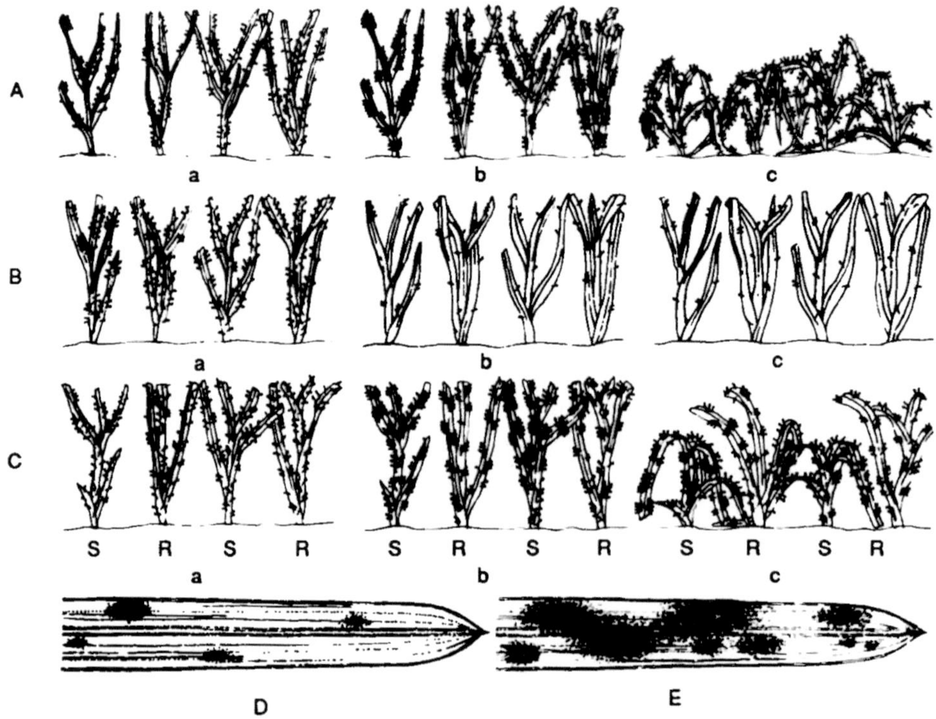

Figure 8.2. Misuse of a blend for disease resistance when good horizontal resistance already is available. *A*, susceptible cultivar: *a*, initial inoculum arrives from susceptible cultivar; *b*, rapid buildup of inoculum in subsequent generations; *c*, susceptible plants dying. *B*, resistant cultivar: *a*, inoculum arrives; *b*, little inoculum is produced in subsequent generations; *c*, resistance by inoculum reduction prevents epidemic. *C*, blend of susceptible (*S*) and resistant (*R*) cultivars: *a*, inoculum arrives on blend; *b*, large buildup of inoculum on susceptible cultivar partially breaks down resistance of resistant cultivar; *c*, combination of susceptible and resistant cultivars yields a turf of intermediate quality. *D*, lesions on resistant cultivar. *E*, lesions on susceptible cultivar.

the Merion with inoculum. The horizontal resistance of Merion may be partly overcome, causing it to express a higher incidence of the disease than if it had been growing as a monostand. Merion may therefore fail to gain the expected competitive advantage over the susceptible cultivar. Evidence for this was found by Vargas and Turgeon.[161]

Does combining horizontally resistant cultivars improve horizontal resistance to a specific disease? So far, the answer is no. Vargas and Turgeon[161] reported that blending horizontally resistant cultivars did not lessen the incidence of disease or improve turf quality.

Stripe Smut

Blending vertically resistant species can prevent multirace host-specific pathogens from destroying a turfgrass stand. Some diseases to which Kentucky bluegrasses have vertical resistance are stripe smut, caused by *Ustilago striiformis*; powdery mildew, caused by *Erysiphe graminis*; and rusts caused by several species of *Puccinia*. Powdery mildew is a problem mainly in the shade, and the easiest way to deal with it is to plant a mildew-resistant grass in shady places. Rusts, with the exception of stripe rust, are a minor problem on slow-growing turf and can usually be managed with adequate moisture and nitrogen. Stripe smut is thus the only major disease of Kentucky bluegrass that requires blending for its management. A combination of three or four Kentucky bluegrass cultivars resistant to current races of stripe smut should provide the broad genetic base required to prevent a race of stripe smut from extensively damaging a stand of turf.

Most of the current stripe smut races are specific for Merion and Windsor. The race that attacks Merion can be designated race 1, and the race that attacks Windsor, race 2. A suitable blend of Kentucky bluegrasses might contain America, Midnight, Able I, and Adelphi, all resistant to races 1 and 2. Suppose that races 1 and 2 hybridize and produce race 3, which turns out to be specific for Adelphi. The vertical resistance of the other three cultivars to race 3 ensures that stripe smut will not ruin the turf. The amount of inoculum produced is not a factor in vertical resistance. Race 3 cannot attack the other cultivars, and the disease affects only the Adelphi plants. However, since resistance to stripe smut is specific, cultivars that are resistant today may be as susceptible as Merion and Windsor tomorrow, as new races of stripe smut develop and become dominant in the population.

Horizontal resistance to stripe smut probably does exist in Kentucky bluegrass, but has been overlooked because the breeding and selecting programs have focused on achieving immunity to stripe smut. What we need to find are cultivars with sufficient general resistance so that no more than 10% or 20% of the plants in a population ever become infected with stripe smut. The same is true for other turfgrass species and other diseases, such as rusts and powdery mildew. Breeding for generalized resistance is much more difficult and usually more time-consuming than is breeding for specific resistance, but the rewards can be much greater.

HIGH-MAINTENANCE TURF

High-maintenance Kentucky bluegrass turfs should consist of cultivars resistant to the four major diseases of Kentucky bluegrass: melting-out, stripe smut, necrotic ring spot, and summer patch. Resistance to melting-out is horizontal, and resistance to necrotic ring spot and summer patch appears to be horizontal too, so blending will do little to improve resistance to those two diseases. Resistance to stripe smut, however, is vertical (at present, anyway), and therefore a blend of three or four cultivars resistant to melting-out, necrotic ring spot, and summer patch is needed to ensure genetic diversity sufficient to prevent a single race of stripe smut from destroying the turf.

Resistance to disease is not the only criterion for cultivar selection. Environmental and cultural adaptation are very important too. A successful cultivar must withstand environmental stress. For example, a cultivar that lacks cold-temperature hardiness cannot survive in the North; resistance to disease will not save it. Cultivars also must be suited to the management regimen under which they will be grown. Kentucky bluegrasses should not repeatedly be mowed to 1/2 inch (1.3 cm), and most cultivars do not grow well in the shade. Fine-leaf fescues will not thrive if irrigation is excessive or drainage is poor. Resistance to disease is no solution to these problems.

Ideally, all cultivars in a blend should meet all three criteria: disease resistance, environmental adaptation, and cultural adaptation. An exception may be made if there is no cultivar known to be horizontally resistant to a multirace pathogen. A single cultivar that meets all three criteria will be superior to a blend containing cultivars that do not meet all three criteria.

CHAPTER 9

The Cool-Season Grasses: A Plant Pathologist's Point of View

Seven principal turfgrasses are grown in the cool season turfgrass growing area: creeping bentgrasses, colonial bentgrasses, Kentucky bluegrasses, annual bluegrasses, fine-leaf fescues, tall fescues, and perennial ryegrasses. In this chapter, I will discuss the strong and weak points of each. I will also include information about where each turfgrass is culturally and environmentally adapted and how it should be maintained.

Although as a plant pathologist I believe the most important characteristic of a grass is its resistance to disease, I am not blind to all other facts. I realize that a turfgrass must have good agronomic qualities, too. The most disease-resistant turfgrass in the world is not practical, desirable, or salable if it looks like pasture grass when it is mowed. On the other hand, a turfgrass with the best agronomic characteristics is just as impractical and undesirable if it is susceptible to a serious disease. Unfortunately, such a turfgrass is salable, as evidenced by the large number of commercially available turfgrasses that are susceptible to serious diseases. This state of affairs is often the fault of university people who list all the agronomic attributes of a turfgrass variety, but fail to mention the diseases to which it is susceptible, mention only the diseases to which it is resistant, or fail to mention anything at all about its disease resistance or susceptibility.

The three turfgrasses for which this problem is most evident are Kentucky bluegrasses, fine-leaf fescues, and perennial ryegrasses. That is not to say that other grasses don't also have problems. Creeping bentgrass, which is grown on golf course greens, is probably the most disease-susceptible turfgrass species of all, but golf course greens are considered high maintenance turf and receive the fungicide treatments necessary to manage or prevent the diseases. Kentucky bluegrasses, fine-leaf fescues, and perennial ryegrasses are grown on areas generally considered to require moderate to low maintenance, such as home lawns, parks, recreation areas, and some low maintenance golf course fairways. No one wants to treat these areas weekly for turfgrass diseases.

To cite a specific example, Prato Kentucky bluegrass is a turfgrass that has very high root density, low growth habit, medium green color, and resistance to powdery mildew. It sounds like a pretty good grass until someone tells you that it is very susceptible to melting-out, stripe smut, and stem rust. (You can usually forget about the stripe smut and rust because Prato is so susceptible to melting-out that it probably won't be around long enough to catch the other diseases.) Another example of a Kentucky bluegrass is South Dakota certified. The word certified has the connotation today of being something special. The only thing it is certified for is the fact that it was grown in South Dakota. Again it is futile to expand on its agronomic characteristics since it is so susceptible to melting-out that a stand of it will be severely thinned and badly invaded by weeds a few years after establishment. A third Kentucky bluegrass cultivar that is susceptible to melting-out is Palouse (Figure 9.1).

In this chapter an attempt will be made to distinguish the desirable turfgrass species and cultivars from the undesirable on the basis of their susceptibility or resistance to the major turfgrass diseases affecting each species. The minor diseases can usually be managed easily by cultural or chemical means. Even though turfgrass diseases are the most important considerations in selecting a turfgrass species or cultivar, a cultivar must also be environmentally and culturally adaptable. Bermudagrass won't do well in Maine, and the best Kentucky bluegrass cultivar won't do well in Florida. Nor can you mow Kentucky bluegrass at 1/4 inch (.6 cm) and expect it to compete with *Poa annua*. It is just as

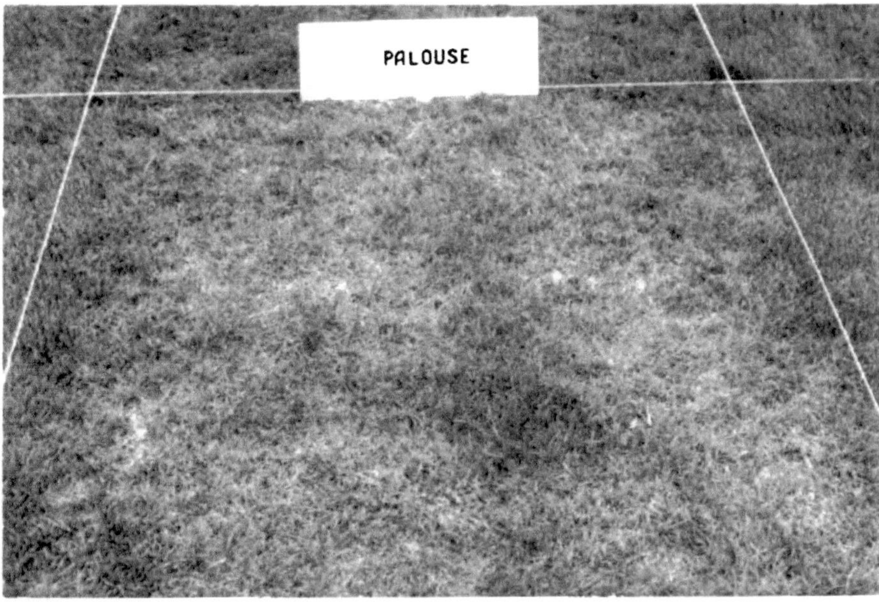

Figure 9.1. Palouse Kentucky bluegrass being severely thinned by melting-out (*Drechslera poae*) during the cool weather of the spring.

foolish to recommend a grass on the basis of its agronomic characteristics, while knowing that a major disease like stripe smut, summer patch, or spring dead spot is going to destroy it.

CULTIVAR EVALUATIONS

There is nothing sacred about cultivar evaluations, and a cultivar that is resistant today may not necessarily be so tomorrow. That is not a "cop-out"—the cultivars recommended in this chapter have the greatest known resistance to the important diseases today. But fungi, like plants, have the ability to change their genetic makeup, and may overcome certain types of resistance. Also, there is the possibility of a new or previously unimportant disease developing or becoming important on a new cultivar. Manhattan perennial ryegrass and red thread provide a good example of this. Before Manhattan was introduced, red thread was not considered an important problem on perennial ryegrass, but now Manhattan is no longer recommended in many areas because of its susceptibility to this disease. Another example is provided by Merion Kentucky bluegrass. It was widely grown because of its excellent resistance to melting-out. After a period of years, stripe smut, which was considered a minor problem on common Kentucky bluegrass, became a major problem on the cultivar Merion. Then necrotic ring spot and summer patch, diseases that were completely unknown before the introduction of Merion, became major problems.

Many turfgrass experts today believe that Merion Kentucky bluegrass has too many disease problems to warrant its recommendation as a desirable turfgrass. A question often asked by the public is "Why, when it is such a poor grass, was Merion so widely used?" The reason that Merion Kentucky bluegrass was so widely grown is its excellent resistance to melting-out. It was the first, and for years the only, Kentucky bluegrass cultivar with such resistance. Perhaps the reason we never saw necrotic ring spot or summer patch as a serious problem in the common-type Kentucky bluegrasses was that melting-out caused severe thinning of the turf and subsequent invasion by weeds which make the symptoms of diseases difficult to discern. The same fate faces every new cultivar released today.

Cultivar evaluations can be made only on the basis of present-day knowledge and must be changed whenever new information becomes available. (The practice of changing cultivars periodically is a hard one for sod growers, landscapers, and others to understand, yet it often is employed by growers of agronomic and horticulture crops.) Even though it is known that the resistance of a cultivar may change and that it may become susceptible, one should use the cultivars that have the best resistance known today in preference to those that are known to be susceptible to major diseases. It makes no sense to start with susceptible cultivars for which there is no hope of success. (For a further discussion of this subject, see Chapter 8.)

THE BENTGRASSES

Creeping Bentgrass (*Agrostis stolonifera* subsp. *stolonifera L.*)

The creeping bentgrasses are some of the most disease susceptible turfgrasses. The species can be considered a desirable one because of its aggressive nature (which allows quick recovery from disease or injury), the fact that it can be maintained as low as $^3/_{16}$ inch (.48 cm) and still survive, and because it is used primarily on golf courses where expensive fungicide spray programs are affordable. In addition to being used on golf course greens, it has become the grass of choice for fairways where the cool season grasses are grown. One reason for its increased popularity has been the advent of clipping collection on fairways which allows the creeping bentgrass to compete better with annual bluegrass. Another reason is people have finally figured out you can't mow Kentucky bluegrass at the ½ inch (1.3 cm) or lower height of cut demanded by today's golfers and expect it to survive or compete with annual bluegrass.

Diseases

The three bentgrass species (creeping, colonial, and velvet) are all susceptible to take-all patch, yellow tufts, pink patch, dollar spot, brown patch, *Helminthosporium* diseases, *Pythium* blight, *Typhula* blight, and *Microdochium* patch. The degree of susceptibility varies among bentgrass species and cultivars, but the variance is slight. The literature is full of discussions of differential susceptibilities of bentgrasses, especially creeping bentgrass, to the various diseases. However, recommendations based on these putative differences aren't worth very much. For example, Penncross creeping bentgrass is listed as moderately resistant to dollar spot.[4] Yet whenever Penncross is grown in an area where dollar spot is a problem, the grass requires fungicide treatments for management of the disease.

Uses

Penncross is the most widely grown creeping bentgrass cultivar on putting green turfs. Penncross, because of its aggressive nature, tends to be grainy which causes inconsistent putting, and requires constant cultivation.

Penncross and Penneagle are suitable as fairway grasses. One must bear in mind that an intensive fungicide program will be required for disease prevention. If you spend the money to establish creeping bentgrass fairways and don't spray them, guess what they are going to become? You guessed it: annual bluegrass fairways.

A brief mention of Toronto creeping bentgrass (C-15) is in order. This vegetatively propagated cultivar used to be the elite bentgrass in the Midwest. It was represented to be genetically homogeneous, but there is now good evidence that at least two distinct strains exist. One strain is susceptible to a disease that has been called "C-15 problem," which was later identified as bacterial wilt. The other strain is not affected by this disease, but it is susceptible to stripe smut. Stripe smut weakens the turf so that it doesn't recover well from traffic, injury, or other diseases, and is replaced by annual bluegrass.

Severe diseases like bacterial wilt and stripe smut have never been a problem on Penncross creeping bentgrass because of the unique way it is produced and the fact it is really a blend of cultivars. The newer creeping bentgrass cultivars are reproduced every year from the same single line cultivars. A single line cultivar continually exposed to a fungal or bacterial population runs the risk of having the makeup of the population change through mutation or selection. It is too bad we recommend blends for every other type of turf stand except for golf course greens and to a lesser degree tees and fairways.

Cultivars

Suggested for Greens

Penncross
Pennlinks
SR 10,20
Putter
Providence

Suggested for Fairways

Penncross
Penneagle
Putter
Providence

Not Recommended

Arlington—Not competitive enough
Congressional—Not competitive enough
Pennlu—Stripe smut
Toronto (C-15)—Stripe smut + bacterial wilt
Washington—Not competitive enough
Emerald—Not competitive enough

Future

Many new cultivars have become available the past few years. Unfortunately, they have not been tested widely enough to make good evaluations. For example, Pennlinks, which has been available now for a few years, does not grow very well in the cool weather of spring and fall and probably should be avoided in the cooler climates. In addition, it is the most susceptible bentgrass cultivar to *Microdochium* patch and yellow tuft I have ever seen. SR 10,20 is super-susceptible to dollar spot, and probably should be avoided on fairways. I am sure we will find other shortcomings of these new bentgrass cultivars as they are more widely grown. You should check with the local turfgrass experts in your area in future years to learn the latest developments concerning these and other cultivars.

Colonial Bentgrass (*Agrostis tenuis* Sibth.)

The colonial bentgrasses are susceptible to a wide range of diseases, from dollar spot to the snow molds. Because of their susceptibility to diseases and lack of a vigorous creeping habit, they are not desirable as turfs for putting greens. They are also not desirable in high-quality turf areas because costly fungicide programs can be avoided by using species that have better disease resistance.

Uses

Do the colonial bentgrasses have a place in modern turfgrass culture? Yes, as a component of mixtures for fairway and general turfgrass areas where, if they die, something else can fill in for them. They survive best in mixtures in the coastal areas of northern Europe and the Pacific Northwest.

Cultivars

The following cultivars are listed, without evaluation, so that you will know what to ask for.

Astoria
Boral
Exeter
Highland
Holfior

Velvet Bentgrass (*Agrostis canina* subsp. *canina* L.)

Velvet bentgrass lacks the disease resistance and competitive ability necessary to make it a desirable turfgrass species for high-maintenance turf areas like golf course greens. Velvet bentgrass is even more susceptible to disease than is creeping bentgrass. In addition, it doesn't wear well, is not very tolerant of high temperatures, and has a tendency to lie flat in the summertime, thus making an undesirable putting surface. In short, because of disease problems and lack of competitive ability, it simply will not compete with annual bluegrass, which eventually will take over. In order to maintain existing velvet bentgrass greens for as long as possible, minimal amounts of nitrogen should be used and the turf should be watered infrequently. A good fungicide program is also necessary.

Cultivars

The following velvet bentgrass cultivar is listed for your information, but no recommendation is given: Kingstown.

THE BLUEGRASSES

Kentucky Bluegrass (*Poa pratensis* L.)

The greatest advances in turfgrass breeding and selection have been made with the Kentucky bluegrasses. There are almost too many Kentucky bluegrasses to name, with new ones arriving on the scene every day. The reason for the explosion in the development of Kentucky bluegrass is simple economics. Most general turf areas in the cool-season turfgrass region are seeded or sodded to Kentucky bluegrasses. For a comparison of the relative amounts of creeping bentgrass and Kentucky bluegrass that commonly are planted, just think of how small an area a golf course green and fairway represent compared to roughs, home lawns, parks, and roadsides. Because of the large acreage involved, not only universities but also many private seed companies have been involved in developing Kentucky bluegrass.

While the burgeoning number of new Kentucky bluegrasses has provided a great variety of colors and textures to choose from, it has also caused great confusion. The buyer wonders which cultivar is the best. The advertisements in the trade journals suggest that they are all super grasses, but this isn't so. The limiting factors in their use, after all is said and done about color, texture, and lower growth habit, are their environmental adaptability, cultural adaptation, and resistance. For no matter how many other good qualities a variety has, if it is not resistant to the major turfgrass diseases and environmentally and culturally adaptable, it won't be around long enough for you to enjoy it.

You can, of course, apply fungicides to manage some turfgrass diseases, but there is an easier way. It is called selecting the right cultivars to start with. The diseases of Kentucky bluegrasses and their avoidance by the proper selection of varieties are discussed below.

Diseases

The easiest way to manage diseases of Kentucky bluegrass is through sound cultural practices and the use of cultivars which offer the best resistance to the major diseases. These major diseases are melting-out, necrotic ring spot, summer patch, and stripe smut (Figure 9.2). The minor diseases of Kentucky bluegrasses are powdery mildew caused by the rusts, and fairy ring.

First, let's dispose of the minor diseases. Powdery mildew is a problem on some Kentucky bluegrass cultivars when they are grown in the shade. It can be avoided by the simple expedient of not planting susceptible varieties in the shade. The rusts are a problem on slow-growing turfs, usually because of a lack of nitrogen. The problem can be eliminated by increasing the amount of nitrogen applied or the frequency of its application, so that the turf is mowed at least once a week. The third minor disease, fairy ring, is really not a disease in the sense of a pathogen attacking a grass host; it is a fungus growing in the thatch or organic matter. The fungus breaks down the organic matter as it grows,

Figure 9.2. A stripe-smut-infected Merion Kentucky bluegrass turf. The infection has caused the grass plants to become clumpy and upright in their growth.

releasing nitrogen which produces the zone of stimulation (the dark green turf associated with the ring). The body (mycelium) of the fungus is hydrophobic and forms a layer that is impervious to water. Consequently, the turf in the area of the main body of the fungus dies from lack of water. The only successful management practices are removal of the fairy ring along with the contaminated soil, and fumigation. Fairy rings are most often found in turf areas where tree branches, roots, or trunks have not been removed, or where they have been used as fill. Avoidance of such practices will help to prevent the development of fairy rings.

Most turfgrass managers do not consider melting-out a major disease problem in Kentucky bluegrass. This is because there are so many resistant cultivars that are available. The main reason Merion was so popular and so widely grown is that, for many years, it was the only cultivar available which had resistance to melting-out. Today, necrotic ring spot, summer patch, and stripe smut receive all the notoriety as the major diseases of Kentucky bluegrass. However, if it were not for the many melting-out-resistant cultivars available, necrotic ring spot, summer patch, and stripe smut would not be as important as they are today. Melting-out would have eliminated the desirable Kentucky bluegrasses long before necrotic ring spot, summer patch, and stripe smut could become a problem.

There are many cultivars with excellent resistance to melting-out, but necrotic ring spot, summer patch, and stripe smut must also be taken into consideration. Merion has excellent resistance to melting-out, but it is very susceptible to stripe smut, necrotic ring spot, and summer patch. Fylking, Nuggett, and Pennstar also have excellent resistance to melting-out, but all are highly susceptible to necrotic ring spot and summer patch. The use of any of these cultivars may result in an unsatisfactory turf.

Kentucky bluegrass cultivars available today that have the best resistance to all three diseases are Adelphi, Midnight, Edmundi, Able 1, Monopoly, America, and Mystic. Since blends give added strength to a turf, especially with respect to such diseases as stripe smut, a blend of three or four of these Kentucky bluegrass cultivars would be ideal. This is not to say that these cultivars will remain resistant, or that some new disease won't come along and destroy them. However, based on our present knowledge, these are the best varieties available. At least there is a chance of having a disease-free turf if one uses these disease-resistant cultivars; no such chance exists when you use such disease-susceptible cultivars as Merion, Fylking, Pennstar, Nuggett, and Windsor.

Uses: General

The improved Kentucky bluegrasses are primarily used on high maintenance home lawns and around commercial buildings in the cooler regions of the cool-season grass-growing area. They are not well adapted for the warmer regions or the transition zone, although people have attempted to grow it there for many years. It does well in golf course roughs when it is mowed between two and three inches (5 to 7.6 cm). It is also not suited for golf course fairway unless you are willing to mow the grass at above an inch.

Uses: Low Maintenance Turf Areas

There are some areas in which the planting of the elite cultivars is not recommended. These are low maintenance areas such as roadsides, parks, and general-use areas where less than 2 pounds of nitrogen per 1000 square feet (1 kg per 100 square meters) per season and no supplemental irrigation are to be applied. In these areas, the common Kentucky bluegrass types should be used. They will survive better under conditions of low maintenance than will the improved Kentucky bluegrass cultivars, which have high nitrogen and irrigation requirements. The turf should be maintained at a minimum 3-inch (7.6 cm) height of cut for best results. Of course, this turf will not have the appearance of a first-class turf composed of properly fertilized and irrigated elite cultivars. It will, however, look and survive better than elite cultivars grown under conditions of low maintenance.

Annual Bluegrass (*Poa annua* L. subsp. *reptans*)

You may have a tough time buying annual bluegrass seed at your local seed store, but don't let that bother you, because annual bluegrass has a way of seeding itself at no expense. I realize that treating annual bluegrass as a desirable turfgrass species is going to be met with amazement in some quarters, because some people still consider it to be strictly a weed. But treating annual bluegrass strictly as a weed is the equivalent of living in the Dark Ages and thinking that the world is flat. Annual bluegrass, or "poa" as it is affectionately known, is a fact of life in the northern areas of the cool-season turfgrass region. It makes up over 80% of the golf course greens and tees over 10 years old, and at least half the fairways. Most chemical-control problems intended to eradicate or prevent the growth of annual bluegrass have failed; many because good annual bluegrass cultural-control programs weren't conducted in conjunction with them. It is about time we faced up to the fact that annual bluegrass can be a good turfgrass species. It is widely grown and, until a satisfactory means of eradication is found, it should be treated as desirable. For a more detailed discussion of all aspects of annual bluegrass, see Reference 6.

Diseases

Annual bluegrass is susceptible to dollar spot, brown patch, *Pythium* blight, leaf spot, summer patch, anthracnose, red thread, *Typhula* blight, and *Microdochium* patch. All of these diseases are important and, if not controlled, will destroy annual bluegrass turfs. Two diseases, anthracnose and summer patch, deserve special mention because they have generally been overlooked as diseases of annual bluegrass (Figure 9.3), and perhaps of the fine-leaf fescues, perennial ryegrass, and some Kentucky bluegrass cultivars as well.[146,149] It has been my experience that what has been called high-temperature killing of annual bluegrass is in reality anthracnose and summer patch. Controlling these diseases, along with the more obvious diseases such as dollar spot, brown patch, and *Pythium* blight, will allow the growing of fine annual bluegrass turfs during the summer in the northern parts of the cool-season turfgrass region. You don't have to like annual bluegrass, but at least you ought to know it dies from diseases like anthracnose and summer patch, and not from high temperature.

Annual bluegrass is resistant to melting-out, stripe smut, powdery mildew, and stem rust.

Uses

Annual bluegrass can be used on golf course fairways and putting greens. The only time it is objectional is in the late spring when it forms many seedheads. An adequate irrigation system is necessary to maintain annual bluegrass in hot weather.

Its resistance to disease gives annual bluegrass a competitive advantage over the Kentucky bluegrasses. When a Kentucky bluegrass turf is thinned by one or more of the diseases to which annual bluegrass is resistant (melting-out, stripe smut, necrotic ring spot), the annual bluegrass seeds that are always present in the soil begin to germinate in the voids (see Figure 9.4).

Annual bluegrass is more competitive than Kentucky bluegrass when a turf is maintained at mowing heights of less than 1 inch (2.5 cm), when nitrogen levels are high, and when irrigation is heavy and fre-

Figure 9.3. Top picture shows anthracnose on treated annual bluegrass fairway 10 days after treatment (on left). Bottom picture shows same fairway 2 months later (left of string); untreated portion of annual bluegrass fairway (right side of string) showing anthracnose damage can be seen in both pictures.

Figure 9.4. Melting-out thinning Kentucky bluegrass during cool spring weather. The annual bluegrass (with seed heads) is not affected by the disease and will probably invade the thinned out areas.

quent. When such management regimens are employed, as they often are on golf course fairways, annual bluegrass will become the dominant species.

Annual bluegrass will also replace creeping bentgrass turf that is damaged by disease. With the exception of anthracnose and summer patch, creeping bentgrass is susceptible to all the diseases that annual bluegrass is. If disease causes an open spot in the turf, annual bluegrass usually will germinate before creeping bentgrass can recover and fill the void. If, on the other hand, annual bluegrass is destroyed by disease, there is enough of its seed in the soil to germinate and replace the destroyed turfgrass when conditions become favorable for doing so.

One final word about annual bluegrass. Before you decide to replace annual bluegrass with something else (and heaven alone knows how you are going to do it without losing your job), ask yourself what you are going to replace it with. You say Kentucky bluegrass. Fine, but you had better get the membership to allow you to mow it to a height of no less than 1 to 1 1/2 inches (2.5 to 3.8 cm), because it won't compete with annual bluegrass if it is cut shorter than that. The fairways should receive as little irrigation as possible, and should be watered only when there is evidence of wilting. (This, of course, means that the members will be playing on hard fairways in the summer.) Finally, applications of nitrogen should be kept to a minimum because excessive amounts encourage the invasion of your Kentucky bluegrass turf by annual bluegrass.

No, you say "creeping bentgrass." Fine again, but now you have at least tripled your fairway fungicide budget. If you don't control leaf spot, dollar spot, brown patch, *Pythium* blight, take-all patch, *Microdochium* patch, and *Typhula* blight, the disease spots in your creeping bentgrass turf will be filled by—you guessed it—annual bluegrass.

Creeping bentgrass fairways must be dethatched mechanically in the spring or they become puffy and scalped in the summer. The cost of mechanical dethatching must therefore be added to your maintenance budget, because annual bluegrass will fill the voids left by scalping. Golfers don't care whether it was scalping or disease that caused the turf to turn brown. To them, brown is brown and they don't like it. Most golfers couldn't tell annual bluegrass from crabgrass, let alone from Kentucky bluegrass or creeping bentgrass, but they can tell green grass from brown, and they usually aren't very fond of brown.

How about the perennial ryegrasses? They are vigorous enough to establish themselves in an existing annual bluegrass turf and are the only alternative available where established annual bluegrass fairways exist. Problem solved? Not quite. Like the other turfgrass species, perennial ryegrasses have their share of disease problems. They are susceptible to brown blight, red thread, crown rust, brown patch, and *Pythium* blight. Furthermore, although the new cultivars mow better than the old common perennial ryegrass, they still don't mow like Kentucky bluegrass, and they have a shabby appearance during the warm summer months unless you keep your mowers extremely sharp.

THE FESCUES

Tall Fescue (*Festuca arundinacea* Schreb.)

In many areas, tall fescue is considered to be a weed, especially when it occurs as single plants, or small groups thereof, in a monostand of a species like Kentucky bluegrass, bermudagrass, or red fescue. However, it can be a desirable turfgrass species for home lawns in the Kentucky bluegrass-bermudagrass transition zone, and for general-purpose use in roadsides and parks in much of the cool- and warm-season turfgrass regions. The newer cultivars have a finer leaf, and when planted in a dense stand provide an excellent turf. Their deep root systems also make them ideal choices for droughty areas that cannot be irrigated on a regular basis.

Diseases

Because of tall fescue's limited use, and the fact that it is treated as a perennial weedy grass, little research has been done on its susceptibility to disease. In general, tall fescue is susceptible to net-blotch and leaf spot, crown rust, *Microdochium* patch, and brown patch. Of all these diseases, brown patch is the most serious disease (Figure 9.5). In fact, it is the limiting factor in the widespread use of tall fescues, particularly on home lawns. There are, undoubtedly, differences in the susceptibilities of different cultivars to these and other diseases; the differences will be discovered as tall fescue is grown more widely as a turf type and as more research is done.

Uses

Tall fescue is used in such areas of low maintenance as roadsides, general park areas, and industrial lawns, and in high-maintenance areas

Figure 9.5. Brown patch on tall fescue (Photograph courtesy of Dr. Robert E. Partyka).

such as home lawns. Because of its wear tolerance, it is also especially well adapted for use on sports turfs and playgrounds in the cool-season grass region. It also can be grown in monostands on home lawns, as it often is in the Kentucky bluegrass-bermudagrass transition zone and in the southwestern United States.

Tall fescue tolerates conditions of low maintenance, especially drought, and it will thrive if nitrogen and supplemental irrigation are applied. It will winter-kill, and it is not really adapted for the extreme northern parts of the cool-season grass region. When it is seeded in polystand with Kentucky bluegrass, the tall fescue should constitute at least 70% of the mixture by weight. The turf will then not exhibit the clumpy appearance associated with tall fescue when it is not the dominant species in a polystand. Tall fescue should not be used on golf course fairways because of its inability to withstand environmental stresses and disease pressure at low cutting height.

The following tall fescue cultivars are listed so that you will know what to ask for if you are ordering one. There is not enough information available on their resistance or susceptibility to diseases to make suggestions. The endophyte content in each cultivar is indicated.

Table 9.1. Endophyte Levels in Some Tall Fescue Cultivars

Adventure (L–N)	Mesa (MH)
Aguara (MH)	Normark 99 (M)
Apache (L–N)	Olympic (L–N)
Arid (M)	Rebel II (L–N)
Falcon (L–N)	Rebel Jr. (M)
Finelawn 1 (L–N)	Titan (H)
Hounddog (L–N)	Tribute (MH)
Jaguar (L–N)	Shenandoah (H)

Key: H - high endophyte
MH - moderately high endophyte
M - moderate endophyte
LN - low or none

The Fine-Leaf Fescues

Chewings fescue (*Festuca rubra* L. subspecies *commutata* Gaud.)
Slender creeping red fescue (creeping red fescue) (*F. rubra* L. subspecies *trichophylla* Gaud.)
Strong creeping red fescue (creeping red fescue) (*F. rubra* L. subspecies *rubra*)
Hard fescue (*F. longifolia* Thuill.)
Sheep fescue (*F. ovina* L.)

Disease remains the limiting factor in the widespread use of the fine-leaf fescues as turfgrass. It used to be thought that leaf spot was the only important disease of the fine-leaf fescues, but some new cultivars have problems with such other diseases as dollar spot, anthracnose, red thread, *Microdochium* patch, and rust. The fine-leaf fescues could well be the grass of the future if it proves possible to incorporate disease resistance into a desirable turf type. They have low water and nitrogen requirements, and as our natural resources become scarcer and more expensive, such species will replace those that require more water and nitrogen if disease resistance adequate for survival without fungicide treatments can be incorporated into them. When fungicide treatment is required to maintain a fine-leaf fescue turf, one natural resource, and its attendant cost, is simply being substituted for another.

Uses

Do the fine-leaf fescues have any uses today? Yes, in areas where low-maintenance turf is used—the same areas, generally speaking, where you would use common Kentucky bluegrass. These are areas where no supplemental irrigation is supplied, and where nitrogen fertility is at a minimum. The fine-leaf fescues' nitrogen requirements are between 1 and 3 pounds of actual nitrogen per 1000 square feet (0.5 to 1.5 kg per 100 square meters) per season. The ideal amount is probably 1 ½ pounds per 1000 square feet (0.7 kg per 100 square meters). The quickest way to get rid of a fine-leaf fescue is to overwater it and overfertilize it with nitrogen. In other words, nothing gets rid of a fine-leaf fescue turf faster than does the installation of automatic irrigation systems. This often happened when the members of a golf course decided that they wanted green fairways in the summer, and that the only way to do that was to install an irrigation system. Once an irrigation system is installed, enough water is usually supplied (overwatering in many cases) to keep the fairways soft and green through the summer. The fine-leaf fescues then begin to disappear, and annual bluegrass begins to take over.

The fine-leaf fescue turfs may go dormant or off-color (brown) in the summer. It is to be hoped that the incorporation of disease resistance into the fine-leaf fescues will make them better able to tolerate water, at least as it relates to disease development, so that green fescue turfs can be maintained through the summer.

The fine-leaf fescues have been placed in such low-maintenance turf areas as roadsides, parks, and golf course roughs. The species is also used in Europe in polystands with colonial or creeping bentgrass for

Table 9.2. List of Suggested Fine-Leaf Fescue Cultivars

Chewings Fescue	
Dover	Lonfellow
Bridgeport (MH)	Southport (HI)
Jamestown II (HI)	Wintergreen
Strong Creeping Red Fescue	
Herald	Seabreeze
Jasper	Vista
Salem	
Hard Fescue	
Aurora (HI)	Reliant
Beljart	Scadlis
Brigade	Serra
Discovery	Warwick
Eureka	
Sheep Fescue	
Bighorn	

Key: HI – High endophyte levels
MH – Moderately high endophyte levels

putting and bowling greens. The fine-leaf fescues are especially well adapted to well-drained, sandy soil and to conditions of low pH (5.5 to 6.5) with little or no added fertility and no supplemental irrigation.

Strong creeping red fescue (*Festuca rubra* subspecies *rubra*) may have the greatest potential for widespread use if a cultivar that has a true creeping habit (spreads laterally by producing rhizomes) can be found. Even though the species is commonly called creeping fescue, 30% or fewer of the plants in any given turf actually have a true creeping habit. At least 70% of the plants must have a true creeping habit if the turf is to be acceptable for general use on golf course fairways.

Cultivars

The recommendations given in Table 9.2 are made with the understanding that the cultivars listed do have disease problems, especially leaf spot, and are not suitable for high-maintenance turf areas.

Meadow Fescue (*Festuca elatior* L.)

Meadow fescue is a tufted species that produces a few short rhizomes. In spite of this, it is not as clumpy as tall fescue when grown in a polystand. In many ways it appears like tall fescue when mowed, although its blades are not as coarse. It tends to blend in better than does tall fescue in a polystand with Kentucky bluegrass. Although it does not have the wear- or heat-tolerance of tall fescue, one cultivar, Beaumont, does have superior direct low-temperature hardiness.

Diseases

Meadow fescue is especially susceptible to net-blotch, brown patch, and crown rust. It is also reported to be susceptible to *Pythium* blight, *Microdochium* patch, and stripe smut. As is the case with tall fescue, cultivars of meadow fescue probably exhibit differing susceptibilities to disease. Information about such differential susceptibilities will become available as meadow fescue becomes more widely grown.

Uses

Meadow fescue is suggested for use in polystands with Kentucky bluegrass in such general-maintenance areas as parks, roadsides, and industrial lawns. It may be useful in polystands with Kentucky bluegrass in sports turfs, especially in northern areas where tall fescue is winter-killed.

Cultivars

There is not enough information available on susceptibility and resistance to disease to warrant recommendations. The following cultivars are listed for purposes of knowing what to ask for, but again, they are not recommended.

Beaumont
Ensign
Minor
Trader

THE RYEGRASSES

Perennial ryegrass (*Lolium perenne* L.)
Italian ryegrass (annual ryegrass) (*Lolium multiflorum* Lam.)

The perennial ryegrasses have become the new kids on the block. They are competing with the Kentucky bluegrasses for use on golf course fairways, home lawns, and in general turf areas. In 1977, an estimated 30 million pounds (13.6 million kg) of perennial ryegrass seed was sold. A great deal of effort, by both universities and private companies, has gone into breeding and selecting new cultivars. It is unfortunate, however, that the major emphasis has been on improving mowability of the perennial ryegrasses and that little attention has been given to disease resistance. Now that a fairly high level of improved mowability exists, perhaps attention can be turned toward developing cultivars with improved disease resistance.

The improved perennial ryegrasses do mow better than common perennial ryegrasses, but they still are not as mowable as Kentucky bluegrass, fine-leaf fescue, or creeping bentgrass. This is especially true during warm weather. Mowing equipment must be kept sharp to prevent shredding, which will result in the turf having an undesirable white cast. It may be difficult to prevent this from happening when rotary mowers are used.

Diseases

The ryegrasses are susceptible to brown blight, crown rust, *Microdochium* patch, brown patch, red thread, stripe smut, *Typhula* blight, and *Pythium* blight, although variation in susceptibility exists among cultivars. Brown blight is an almost universal problem on the ryegrasses; it occurs in the cool weather of spring and fall. In the northern parts of the cool-season turfgrass region, the perennial ryegrasses that aren't winter-killed are severely injured by *Microdochium* patch and *Typhula* blight. In the coastal areas of North America and Europe, where cool weather in the spring and fall slows the growth rate of turfgrass, crown rust and red thread are serious problems. *Pythium* blight is a problem in the warm-season turfgrass areas, where perennial ryegrass is used to overseed dormant bermudagrass, and in the transition and cool-season zones, where the new improved perennial ryegrasses are being used as fairway grasses (see Figure 9.6).

The perennial ryegrasses are the most susceptible of all the desirable turfgrasses to *Pythium* blight, brown patch, red thread, and crown rust. There are no real differences among cultivars in their susceptibilities to these diseases, and they all require fungicide treatment. But other factors, such as heat tolerance, cold tolerance, and competitive ability, must also be considered when one is selecting a perennial ryegrass cultivar. There has not been enough research conducted on the cultural requirements of ryegrass under different maintenance regimens to make specific cultivar suggestions possible.

Perennial ryegrasses do have one advantage over most grasses at this time in that many of them have endophytes. These are fungi that live in a mutualistic symbiotic state with the ryegrass plants, and protect them from foliar feeding insect pests. These endophytic fungi that are found

Figure 9.6. ***Pythium*** **blight on a perennial ryegrass lawn (Photograph courtesy of Dr. Bobby G. Joyner).**

in the aboveground portions of plants; i.e., leaves, stems, and seeds; produce a toxin that, when ingested during feeding, kills the insect. Perennial ryegrass plants that contain endophytes also seem to survive the summer stress period better.

The selection of cultivars listed in Table 9.3 has been made simply on the basis of their content of endophyte.

Uses

Italian ryegrass is used primarily as a nurse grass where quick establishment is needed prior to the germination and establishment of the permanent grasses. This is especially the case where slow-germinating

Table 9.3. Perennial Ryegrass Cultivars with High Endophyte Levels

Yorktown III	Prelude II	Legacy
Palmer II	Repell II	4 Del. Dwarf
Gen–90	Assure	Pinnacle
Express	Pleasure	Repell
Advent	Target	SR 4200
Seville	Riviera	Commander
Dandy	Gettysburg	Regal
Duet	Pennant	Saturn
Manhattan II		

Kentucky bluegrasses are used on erosion-prone areas. Unfortunately, they germinate so quickly and are so coarse that they often out-compete and shade the Kentucky bluegrass to the point that when the Italian ryegrass dies, there is very little Kentucky bluegrass. You are better off hydro-seeding or using some other type of soil stabilization procedure rather than using Italian ryegrass.

One of the major uses of perennial ryegrass is the overseeding of bermudagrass greens when they go dormant in the fall. The perennial ryegrasses supply a desirable green color and, probably more important in the long run, they protect the bermudagrass from excessive wear while it is dormant. The seeding rate used is around 30 pounds per 1000 square feet (15 kg per 100 square meters)! Dormant bermudagrass fairways on some golf courses also are overseeded to improve their color, but a much lighter rate of seeding is used.

Many golf course superintendents who have given up trying to grow bermudagrass and Kentucky bluegrass in the transition zone between the warm- and cool-season grassbelts have turned to perennial ryegrass. They may be sitting on powder kegs. *Pythium* blight is difficult to control on perennial ryegrass fairways during hot, humid weather. It requires many applications throughout the summer. Care should be taken to apply low-rate mixtures of metalaxyl, propamocarb, and fosetyl-Al to prevent resistance from occurring. See Chapter 6 for detailed discussion of preventing fungicide resistance.

The perennial ryegrasses are also used on sports turf. Since they germinate so quickly, they are excellent for seeding worn areas. Mowing quality may be poor, and diseases such as brown blight, crown rust, and red thread can be a problem, especially in the fall when sports fields are used for football. However, there really are no choices for such heavily trafficked areas.

The perennial ryegrasses are excellent grasses for tees, especially par three and practice tees, where divots are a major problem. The perennial ryegrasses germinate quickly in 3 to 5 days, and they germinate all season long, even during the warm weather of summer. They also do well on shaded tees, especially where tree root competition is a problem.

CHAPTER 10

The Warm-Season Grasses: A Plant Pathologist's Point of View

THE ZOYSIAGRASSES (*ZOYSIA* WILLD.)

Now that zoysiagrass has been used more extensively, its disease problems are coming to light. In particular, zoysia patch has caused severe losses to zoysiagrass turfs. Management of this disease is very expensive, and the results to date have been very inconsistent. Zoysiagrass is slow to establish itself, but it eventually forms a dense turf that is practically weed-free. Zoysiagrass is also slow to recover from injury, which may be an undesirable characteristic for golf course fairways and tees. Zoysiagrass does have great potential for general turfgrass areas, as well as for golf course fairways in the cooler portion of the warm climatic zone and in the bermudagrass-Kentucky bluegrass transition zone. Its cultural requirements, disease resistance, and susceptibility still require further study.

While zoysiagrass has great potential for the warm climatic zone, it is not well adapted to cool temperatures. It does have winter hardiness that allows it to survive, but it discolors (browns) with the first cool weather (approximately 55°F [13°C]) of the fall, and does not green-up until late spring or early summer. This means that anyone who plants zoysiagrass in the cool-season grass region can expect a green turf for 2 to 3 months per year and a brown turf for 9 to 10 months. In spite of this difficulty, large advertising campaigns that describe the slow-growing, less-mowing, weed-free, dense-turf virtues of zoysiagrass appear every year in large metropolitan areas. The advertisements don't mention the fact that zoysiagrass is slow to green-up and becomes off-color early in the fall.

Diseases

Zoysiagrass is reported to be susceptible to dollar spot, brown patch, rust (*Puccinia zoysia*), and *Helminthosporium* disease. However, all of these diseases except brown patch are minor and for the most part can be managed culturally. Brown patch is a serious disease, particularly on golf course fairways, and does require fungicide treatments. Zoysia patch, on the other hand, is a devastating disease and has begun to threaten the widespread use of zoysiagrass as an acceptable turfgrass species.

Uses

Zoysiagrass can be used on home lawns, parks, golf course tees, and fairways, and as a border grass on approaches to prevent the bermudagrass on the fairway from invading the bentgrass green.

Cultivars

There are several improved zoysiagrass cultivars. They are listed below for informational purposes, but no recommendation is made.

Korean Common
Meyer
Emerald
El Toro
Belair

CENTIPEDEGRASS [*EREMOCHLOA OPHIUROIDES* (MUNRO) HACK.]

Diseases

Brown patch, dollar spot, centipedegrass decline, and nematodes can cause severe injury.

Uses

Centipedegrass is grown mostly on low-maintenance turf areas. It will not persist where soil pH is high or where high concentrations of nitrogen and excess water are applied. Centipedegrass is used on home lawns or other areas where minimal care is given and there is little traffic. Like the fine-leaf fescues and certain Kentucky bluegrasses, it does quite well under such low-maintenance conditions. Disease problems, as well as the direct effects of high maintenance, may be responsible for the loss of centipedegrass under intense culture. Oklawn is one improved cultivar of centipedegrass that is available, although seed supply is limited. It was released by the Oklahoma Agricultural Experiment Station, and is apparently well adapted for that area. No disease information is available.

BUFFALOGRASS [*BUCHLOE DACTYLOIDES* (NUTT.) ENGELM.]

Buffalograss grows naturally in many mowed, unirrigated areas of the warm arid climatic zone. It persists in unirrigated fairways and roughs on golf courses. Very little selecting or breeding has been done with this species, but as natural resources become more scarce it could become an important species, especially in arid and semiarid areas of the warm arid climatic zone.

Diseases

There is no information available on the susceptibility of buffalograss to disease.

Uses

Buffalograss is the fine-leaf fescue of the arid and semiarid warm climatic zone. It survives well in areas where no supplemental irrigation and little or no supplemental nitrogen are applied. Irrigation and the application of nitrogen may cause buffalograss to decline.

Cultivars

Prairie
609
Commanche
Buffalawn
Sharp's Improved
Bison
Texoka

ST. AUGUSTINEGRASS [*STENOTAPHRUM SECUNDATUM* (WALT.) KUNTZE]

St. Augustinegrass, like most of the warm-season grasses, is currently propagated only vegetatively. It has been widely used on home lawns, industrial grounds, and parks. It is not low temperature hardy and is restricted mainly to the southern part of the warm humid climatic zone.

Diseases

St. Augustinegrass is susceptible to brown patch, especially in the shade (Figure 10.1), gray leaf spot, downy mildew, dollar spot, and St. Augustine decline virus (SADV). A resistant cultivar of St. Augustinegrass was found by some fine cooperative efforts between the University of Florida and Texas A&M University. This resistant cultivar is called Floratam. Unfortunately, while Floratam has good resistance to SAD, it lacks good turfgrass quality in terms of leaf texture. It also lacks shade adaptation and, in comparison to Texas common, it lacks low-temperature hardiness.

Cultivars resistant to SAD

Raleigh
Serville
Foralawn
Better blue

Figure 10.1. Brown patch on St. Augustinegrass (Photograph courtesy of Dr. Don Blasingame).

Other cultivars

Jade
Mercedes
Sunclipse
Delmar

BERMUDAGRASS (*CYNODON* SPP., L.C. RICH)

Bermudagrass is one of the most important warm-season turfgrasses. It has a wide range of uses, from golf course greens, tees, and fairways to home lawns, athletic fields, institutional grounds, cemeteries, roadsides, and general park areas. A few of the improved cultivars can be mowed as low as 1/4 inch (.6 cm), although heights of 3/4 to 1 inch (2 to 2.5 cm) are preferred for general purpose use. One of the things that makes bermudagrass so popular is its recuperative capability after it has been thinned by traffic, disease, or insects. Bermudagrass does have poor low-temperature tolerance. Also, it loses color in the cool weather (55°F [13°C]) of the fall and doesn't green-up until soil temperatures rise to 64°F (18°C) at a 4 inch depth in the spring.

Diseases

Bermudagrass is susceptible to spring dead spot (Figure 10.2), *Helminthosporium* diseases, brown patch, dollar spot, *Pythium* blight, rust, and *Microdochium* patch. Spring dead spot (SDS) is the limiting factor in the successful growth of bermudagrass as a turfgrass species in many areas of the warm climatic zone. The colder the winter, the more severe is spring dead spot. The disease is an especially severe threat to bermudagrass in the transition zone and in cooler areas of the warm climatic zone.

Bermudagrass golf greens that are on a heavy fungicide spray program very seldom have a problem with spring dead spot. Wilcoxen[168] showed that repeated fungicide application on fairways during the growing season reduced the amount of SDS. While more research needs to be conducted on chemical management of SDS, it may be worth trying such a program where quality bermudagrass with a reduced incidence of spring dead spot is desired.

Bermudagrass is a vigorous turfgrass that responds well to nitrogen. It does best when adequately irrigated.

Bermudagrass has more diseases listed for it than do the other warm-

Figure 10.2. Spring dead spot (SDS) on a bermudagrass golf course fairway (Photograph courtesy of Dr. Bill Daniels).

season grasses. This is due in part to the fact that it has been studied more, and also because it has been used on golf courses more than any other warm-season species. It also has been subjected to high-intensity cultural practices more than has any other warm-season grass, and this has led to the discovery of its disease problems.

Cultivars

The cultivar Midiron is currently being represented as resistant to spring dead spot. If it has resistance, or if this resistance holds up, it will be a major breakthrough and will greatly expand the area in which bermudagrass can successfully be grown. It should also supply the genetic material to start a breeding program to develop other cultivars with resistance to SDS.

Four species of bermudagrass are grown as turfgrasses: *Cynodon dactylon* (L.) Pers., *Cynodon transvaalensis* Burtt-Davy, *Cynodon magennisii* Hurcombe, and *Cynodon incompletus* var. *hirsutus* (Stent) de Wet et Harlan. Several of the improved cultivars are listed below. They may be used in areas where spring dead spot is not a problem or where fungicide programs are used.

Figure 10.3. Lawn severely thinned by nematodes (Photograph courtesy of Dr. Don Blasingame).

Dactylon bermudagrasses
- Nu Mex Sabora
- Sonesta
- Flora Tex
- U-3

Hybrid bermudgrasses
- Midiron
- Midway
- Santa Ana
- Tex turf 10
- Tifdwarf
- Tifgreen
- Tifway

OTHER PROBLEMS OF WARM-SEASON TURFGRASSES

Insect and nematode problems (Figure 10.3) are more severe in warm climates than they are in cool ones. This is not to say that the cool-season grasses don't have nematode or insect problems, but these problems are more numerous and occur more frequently on the warm-season grasses.

CHAPTER 11

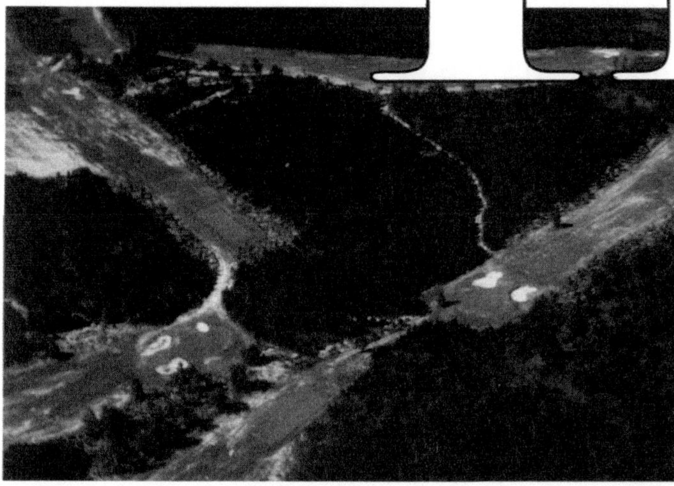

Disease Management Strategies for Golf Courses

COOL-SEASON GRASSES: GREENS

The Bentgrasses

Suggested Species

Let us start with the easy one, turfgrasses used for golf course greens. You are talking about creeping bentgrass. Yes, some people still plant velvet bentgrass, and if you want a grass that is extremely susceptible to common bentgrass diseases, won't compete well with annual bluegrass, becomes unmanageably puffy in hot weather, and provides one of the poorest-quality putting surfaces ever, please don't let me discourage you from planting velvet bentgrass.

Creeping bentgrasses can be assigned to one of two classes on the basis of the way in which they are established: there are those that can be established only vegetatively, and those that can be established either vegetatively or from seed. Toronto (C-15) and Cohansey (C-7) were the two most successful vegetatively established cultivars. Neither of these cultivars is available today.

The most widely adapted and most widely used of the seeded cultivars is Penncross creeping bentgrass. It has good wear tolerance[5] and is vigorous enough to compete fairly well with annual bluegrass if diseases are controlled and proper cultural practices are followed. Pennlinks, Providence Putter, and SR-10,20 are becoming popular for putting green turfs. They are less grainy and have finer leaf blades, which provides for a better putting surface. However, none of these cultivars has been grown long enough to determine if they have major problems. For example, we are just beginning to notice that Pennlinks tends to go off-color in the cooler weather and may not be a desirable grass in the cooler regions of the world. The overall disease susceptibility of these cultivars is about the same, although SR-10,20 is very susceptible to dollar spot, and Pennlinks is very susceptible to yellow tuft and *Microdochium* patch. To evaluate the other seeded types of creeping bentgrass that have been grown widely—Emerald, Seaside, and Penneagle—none of them are vigorous enough to make a good putting surface and compete with annual bluegrass.

Diseases

Creeping bentgrasses are susceptible to take-all patch, dollar spot, brown patch, *Pythium* blight, *Typhula* blight, *Microdochium* patch, pink patch, leaf spot, crown-rotting anthracnose, and red leaf spot.

Cultural Programs

Irrigation. The greens are obviously the most important single part of a golf course. A few bare spots on the tees and fairways may go unnoticed, but they never will be unnoticed on a green. In order to keep the greens free of bare spots, an adequate watering system and good surface and subsurface drainage are essential. Standing water in hot weather can be as damaging as too little water. When purchasing an irrigation system, get a good one. It will be cheaper in the long run to spend a little extra money to do it correctly rather than to nickel-and-dime yourself to death trying to correct problems after the fact with stopgap measures that only make a bad situation temporarily acceptable.

Golf course greens should be watered as needed to prevent wilting. The amount of water used will depend on such things as soil type, time of year, and species of grass. Remember that creeping bentgrass greens should be kept on the dry side to prevent the invasion of annual bluegrass. Light daily irrigation regimes that do not replace the oxygen in the root zone have proved very beneficial in maintaining heathy creeping bentgrass greens. This has proved especially effective in maintaining creeping bentgrass greens in warm-season grass-growing areas like the southern U.S. During warm weather, greens are often only syringed with the misting sprinklers during the day. This has the dual effect of cooling the plants off during the heat of the day while not applying excess water to the root zone, which will push oxygen out and cause root deterioration.

Mowing Height. Creeping bentgrass greens can be mowed between 1/8 and 5/16 inch (.3 and .8 cm). They *should* be mowed between 3/16 and 1/4 inch (.5 and .6 cm); unfortunately, today's golfers demand putting speeds that require greens to be mowed around 1/8 of an inch (.3 cm). It is very difficult to maintain creeping bentgrass at that height. Creeping bentgrass greens mowed at 1/8 inch (.3 cm) or less quickly become annual bluegrass putting greens! For all the fancy explanations of overwatering and overfertilizing being responsible for the loss of creeping bentgrass and the encroachment of annual bluegrass, nothing will cause the loss of creeping bentgrass from a green and the encroachment of annual bluegrass faster than an 1/8 inch (.3 cm) mowing height. It is simply too short a mowing height for creeping bentgrass to survive, and is more suitable for annual bluegrass.

Some cultural and chemical practices that may allow you to mow your greens higher and still maintain acceptable speeds are light frequent sand topdressing, double cutting, occasional rolling, and low-rate applications of a plant growth regulator. If your goal is to have creeping bentgrass greens, these are viable alternatives to lowering the mowing height for increased putting speed.

Fertility. Excessive use of nitrogen is one way to lose your grass in a hurry. To most people, a healthy turf is one that is dark green. A healthy turf is one that has good shoot density; it may or may not be dark green. As more nitrogen is supplied, more shoot growth is produced, but this is usually at the expense of the root system. Nitrogen causes the foliar tissue to become more succulent, and the more succulent it becomes, the more susceptible it becomes to serious diseases such as *Pythium* blight and brown patch. Too much nitrogen also will make golf course greens more susceptible to wilting, desiccation, and wear.

There is no pat formula for determining the amount of nitrogen that should be applied, and that is why growing grass is an art as well as a science. The amount of nitrogen that is optimum varies with the length of the growing season, soil type, amount of watering, and other factors. However, there is a general rule of thumb for Penncross and Pennlinks creeping bentgrass: no more than 3 pounds of nitrogen per 1000 square feet (1.5 kg per 100 square meters) per season should be applied in the cooler regions of the cool-season turfgrass belt, and no more than 4 pounds per 1000 square feet (2 kg per 100 square meters) should be applied in the southern regions. These are maximum permissible amounts, and less nitrogen should be used where possible.

If a dark green color is desired, use iron, because trying to obtain a dark green color with nitrogen will only lead to a lot of grief. Adequate levels of phosphorus and potassium should be maintained. Amounts to apply should be determined by a soil test. The phosphorus level should be between 40 and 70 pounds per acre (46 and 80 kg per hectare), and the potassium level should be between 175 and 250 pounds per acre (196 and 280 kg per hectare).[114]

Cultural Practices. Cultural practices that must be carried out on a golf course green if a quality putting surface is to be maintained include vertical mowing, brushing, and combing, or, to remove graininess, all three. Topdressing should be done to reduce thatch and provide a firm putting surface. Spiking, usually weekly, will help break up surface compaction, and coring one to three times a season during cool weather will relieve compaction at the 1–3 inch (2.5 to 7.6 cm) depth, where it is a problem.

Shade. Creeping bentgrass will not survive at greens height without morning sun. It does fairly well until late July or early August when the sun begins to rise lower on the horizon. It doesn't matter how much afternoon sun the greens receive, if it doesn't get morning sun, the creeping bentgrass begins to thin and die. There are only two choices: cut down the trees, which is almost sacrilegious at most courses, or collect the clippings from your (of course, you don't have any) or someone else's annual bluegrass fairways and seed the greens to annual bluegrass. Whether you seed them initially to annual bluegrass or slowly watch them go to annual bluegrass is immaterial; the fact is they will eventually become annual bluegrass greens. It is just that it might be better to have healthy annual bluegrass greens from day one instead of suffering through 5 to 10 years of watching unhealthy creeping bentgrass greens become annual bluegrass greens.

Annual Bluegrass

I know that this sounds blasphemous, but the simple fact is that annual bluegrass exists on the majority of the golf course greens in the cool-season grassbelt and, since there is no effective, safe means of eliminating it, replacing it with something else, or preventing it from coming back, it must at least be classified as a functionally desirable species. Certainly it is preferable to bare ground. If you don't think so, talk to a superintendent who has successfully eliminated annual bluegrass by chemical means, or lost it for some other reason, and ask him how he begged and prayed that the "lousy, miserable, good-for-nothing annual bluegrass" would come back and cover up the bare ground. The point being made here is that before you eliminate the annual bluegrass, you had better know how much you really have, not how much you tell everyone you have. If annual bluegrass constitutes more than 20% of your turf, ask yourself what is going to be in those spots when it dies. If the answer is dead dirt, you may wish to think twice before you eliminate the annual bluegrass.

Before you can maintain annual bluegrass properly, you must first admit that you have it. The three greatest lies told by a golf course superintendent to his fellow superintendents are: "My greens chairman never interferes," "I get everything I ask for," and "I have hardly any annual bluegrass in my greens." First you have to face the facts and admit that you have annual bluegrass greens; then you can use a maintenance, fertility, and pest-control program designed for annual bluegrass and not for creeping bentgrass greens.

Diseases

Annual bluegrass is susceptible to dollar spot, brown patch, *Pythium* blight, crown-rotting anthracnose, anthracnose, *Microdochium* patch, *Typhula* blight, leaf spot, and summer patch (Figure 11.1).

Cultural Programs

Irrigation. Contrary to popular belief, annual bluegrass will root as deeply as will colonial bentgrass, creeping bentgrass, or Kentucky bluegrass in similar soils.[169] Light, daily irrigation that doesn't eliminate oxygen from the root zone is beneficial to healthy annual bluegrass roots. This is especially important in reducing the severity of summer patch. Daily irrigation during the warmest part of the day, instead of during nighttime hours, also helps cool the grass plant so it can survive

Figure 11.1. Annual bluegrass fairway damaged by anthracnose, with healthy fungicide-treated annual bluegrass approach and green in the background.

the heat stress periods of summer. Summertime syringing, during the warmest part of the day, may be necessary where nighttime irrigating regimes are used.

Mowing Height. Annual bluegrass, like creeping bentgrass, should be mowed to between $^1/_8$ and $^5/_{16}$ inch (.3 to .8 cm). However, unlike creeping bentgrass, it can tolerate the $^1/_8$ inch (.3 cm) mowing height. In fact, there is little choice for anything other than annual bluegrass or at least annual bluegrass-creeping bentgrass mixtures on greens that are mowed at $^1/_8$ inch (.3 cm) height. Mowing twice daily in the spring when heavy seedhead production occurs will help maintain a good putting surface.

Fertility. Annual bluegrass loves nitrogen. More than that, it requires nitrogen if it is to thrive, especially during the summer. The phosphorus and potassium should be maintained at adequate levels based on soil tests (see Chapter 7).

Cultural Practices. Vertical mowing to remove graininess is not as necessary with annual bluegrass as it is with creeping bentgrass greens, but light vertical mowing will help to reduce some of the adverse effects on putting that are caused by heavy seedhead formation in the spring. Brushing or combing also will help to maintain a truer putting surface. Coring one to three times a season, preferably following heavy seedhead formation will aid in new root production.

COOL-SEASON GRASSES: TEES

Suggested Species

Creeping bentgrass and perennial ryegrass should be used if any grass other than annual bluegrass is desired on a golf course tee. Penncross creeping bentgrass is the preferred cultivar because of its vigorous growth habit. There are many perennial ryegrasses to chose from for golf tees. (See Chapter 10 for choices.)

Penncross tees should receive the same treatment as Penncross greens. They should be cored, vertically mowed, and topdressed. Otherwise, compaction will take place in trafficked areas, and puffiness, resulting in scalping during warm weather, will occur in areas of the tee that are not extensively used. Tees should also provide firm footing for the golfer; they will not do so unless proper cultural practices are followed. A good preventive fungicide program is necessary to prevent the loss of the Penncross turf to diseases, and the subsequent invasion of annual bluegrass. The Kentucky bluegrasses and the fine-leaf fescues are not suitable for golf course tees, in spite of the fact they are still used.

Cultural Programs

Irrigation. An irrigation program appropriate for creeping bentgrass greens (see pages 216–217) is also suitable for creeping bentgrass tees, and the same is true for annual bluegrass (see pages 219–220).

Fertility. The fertility requirements of creeping bentgrass or annual bluegrass tees are the same as those of the greens (see pages 217–218 and 220).

Cultural Practices. Creeping bentgrass tees should receive the same cultural program as creeping bentgrass greens. In addition, the divots, especially on par-3 holes, should be filled periodically throughout the growing season with a soil that contains seed of the grass species you want to have on your tee. If this grass is annual bluegrass, then it should suffice simply to fill the divots with soil.

COOL-SEASON GRASSES: FAIRWAYS

There are five cool-season grasses that often are intentionally used on golf course fairways: creeping bentgrass, colonial bentgrass, Kentucky bluegrass, fine-leaf fescue, and perennial ryegrass. Another grass, which is not planted intentionally but makes up 70% to 80% of most golf course fairways that are over 10 years old, is annual bluegrass. This is not to say that all fairways consisting of other species are doomed to become annual bluegrass fairways, but it is inevitable that they will, unless strict cultural practices are followed.

The four key factors for the survival of Kentucky bluegrass and fine-leaf fescue are disease control, mowing height, irrigation, and fertility levels. The minimum height of cut for these species is 1 inch (2.5 cm), with 1 1/2 inch (3.2 cm) preferred. If they are cut shorter they will not compete with annual bluegrass. (However, you should note that most golfers want the fairways mowed to 1/2 inch (1.3 cm).) But for certain low-budget golf courses where minimal irrigation and fertility are used, Kentucky bluegrass and fine leaf fescue may survive in mixed stands with annual bluegrass.

The Bentgrasses

Recommended Species

Colonial bentgrasses can be used as a minor component of mixtures with Kentucky bluegrass or fine-leaf fescues, but by themselves they simply do not make suitable fairway turfs.

Creeping bentgrass can be used as a fairway turf, and it will compete with annual bluegrass when mowed at 1/2 inch (1.3 cm). In order to give the creeping bentgrass a competitive advantage over annual bluegrass, it should be irrigated as little as possible, grown where there is good drainage, and have nitrogen fertility kept at a minimum. The creeping bentgrasses have little disease resistance and are susceptible to dollar spot, *Pythium* blight, brown patch, leaf spot, *Typhula* blight, *Microdochium* patch, and many other diseases. They must thus be treated with fungicides throughout the year. If creeping bentgrass turfs are not so treated, annual bluegrass will fill in the bare spots and thinned areas left by the various diseases. Furthermore, unless some type of vertical mowing is practiced, creeping bentgrass will become puffy during hot weather and scalping will occur. This will lead to the invasion of annual bluegrass in the scalped areas during the fall. The necessary disease management and cultivation practices will add considerable expense to a maintenance budget—a fact that should be taken into consideration before planting fairways with creeping bentgrass.

Diseases

The same diseases that attack creeping bentgrass on greens and tees also attack it on fairways. However, many people forget this fact when they put in creeping bentgrass fairways—they forget to multiply the $3,000 fungicide budget for their 2 to 3 acres (about 1 hectare) of greens and tees by a factor large enough (at least 10) to keep their creeping bentgrass fairways disease-free. It should, of course, go without saying that if you don't keep them disease-free, then it won't be long before the creeping bentgrass fairways will become "creeping" annual bluegrass fairways.

Cultural Programs

Irrigation. Irrigation should be done lightly on a daily basis. Fairways should receive between 0.1 (0.25 cm) and 0.4 inch (1 cm) of water, depending on the evaporation rate. About 50% to 60% of water loss from an open evaporation pan should be replaced on a daily basis.

Mowing Height. Creeping bentgrass fairways should be mowed to a height of 5/8 inch (1.6 cm) or less.

Fertility. Nitrogen levels in the soil should be kept to a minimum. Excessive nitrogen will only increase susceptibility to diseases and environmental stress, reduce wear tolerance, and increase the amount of thatch. The latter effect could lead to serious scalping problems. High nitrogen levels will also favor the encroachment of annual bluegrass. Adequate phosphorus and potassium levels should be maintained, based on yearly soil tests.

Cultural Practices. Creeping bentgrass fairways need all the cultural practices that creeping bentgrass greens and tees receive except topdressing, and they really need that; it is just that the cost is prohibitive. Coring should be done at least once a year, and should be followed by dragging with a heavy steel mat to break up the cores and incorporate the soil into the thatch. The turf should then be swept to remove surface debris. Fairways should be vertically mowed two or three times in a season, and this should be followed by sweeping to collect the debris. This practice will help to keep the thatch at a reduced and manageable level. It will also reduce the puffiness associated with bentgrass fairways and prevent scalping during hot weather.

Annual Bluegrass

You can't buy it anywhere. Nobody admits they have it. And yet it makes up 70% to 80% of golf course fairway turfs that are over ten years old in the middle and cooler regions of the cool-season turfgrass belt. Some people have successfully decreased the amount of annual bluegrass in their fairways, most have not, many have lost their jobs trying to get rid of it, and the smart ones have learned to live with it. We are talking about "poa"—*Poa annua*, or annual bluegrass.

Annual bluegrass makes a fine fairway turf. It can take a one-half inch (1.3 cm) mowing height, and affords the golfer an upright lie. It does not, as was believed for many years, die from high-temperature stress in the northern and middle regions of the cool-season grass belt. It will die in the summer during periods of heat stress, if diseases and insect problems are not controlled.

The basic problem one encounters in growing annual bluegrass is the lack of information about its cultural requirements and pest problems. It was long treated as a weed, and little research was done on it. People, therefore, tried to adapt management practices appropriate to Kentucky bluegrass and creeping bentgrass to the growing of annual bluegrass. This has resulted in two big mistakes. First, anthracnose, summer patch and the black turfgrass *Ataenius* beetle are not problems on Kentucky bluegrass and creeping bentgrass (at least in the middle and cooler regions of the cool-season grassbelt), but they are problems on annual bluegrass (see Figure 11.2). Die-off caused by anthracnose, summer patch, and the *Ataenius* beetle was wrongly attributed to the effects of high temperature. Secondly, although creeping bentgrass will survive without summer fertilization (in some cases the application of excess nitrogen during the summer actually leads to severe disease problems in this turfgrass), this is not the case with annual bluegrass. Annual bluegrass requires light applications of nitrogen fertilizer (one-half pound per 1000 square feet (.25 kg per 100 square meters) per month) to best survive the summer. More research needs to be done on annual bluegrass and its unique requirements, but to maintain it in the

Figure 11.2. Fairway thinned by black turfgrass *Ataenius* beetle. The areas of damage are well defined, whereas damage caused by anthracnose is more general and usually affects the entire fairway.

meantime one must stop using cultural and pest-control programs that were developed for Kentucky bluegrass and creeping bentgrass.

Pre- and post-emergence herbicides notwithstanding, there is now no satisfactory chemical means of safely eradicating annual bluegrass or of preventing it from invading a green, tee, or fairway. The ultimate solution may be more-aggressive bentgrass that will crowd out the annual bluegrass. For now, the good cultural practices and the use of plant growth regulators that favor creeping bentgrass over annual bluegrass should be used. This will probably work on sandy or sandy loam soils and where there is full sunlight. It may be futile to try to prevent invasions by annual bluegrass on the really fine-textured, heavy soils, especially under shady conditions.

Diseases

Annual bluegrass is susceptible to dollar spot, leaf spot, brown patch, *Pythium* blight, red thread, anthracnose, *Typhula* blight, summer patch, and *Microdochium* patch. It's summer patch and anthracnose, and not the heat alone, that kills annual bluegrass during hot weather. Annual bluegrass also has problems with two insect pests. One of these is a small grub called black turfgrass *Ataenius* (*Ataenius spretulus*); the other is the turfgrass weevil, which causes annual bluegrass to thin and die during hot weather.

Maintenance Programs

Irrigation. Annual bluegrass should be irrigated lightly every day. Syringing may be necessary to maintain annual bluegrass during warm weather.

Mowing Height. Annual bluegrass should be mowed to a height of $^1/_2$ to $^5/_8$ inch (1.3 to 1.6 cm).

Fertility. One half pound per 1000 square feet rates of nitrogen per month will maintain healthy annual bluegrass. Excess nitrogen that leads to lush growth should be avoided, because it favors the development of disease. Adequate levels of phosphorus (P_2O_5) and potassium, based on soil tests, should be maintained for healthy annual bluegrass fairways.

Cultural Practices. Coring once or twice a year should be all the cultivation that is required on annual bluegrass fairways. The best time to core is following heavy seedhead formation in the spring. When annual bluegrass goes through heavy seedhead formation, most of the plant's energy is diverted into seedhead formation. New root growth ceases, and it has been estimated that over 70% of the root system that is present stops functioning. Following this period it is important that the annual bluegrass plant develop as good a root system as possible to aid it through the summer stress period. Coring to provide holes for excellent root growth is important. Remember, cultural practices should be aimed at developing as strong a plant as possible to allow it to survive the summer stress period.

Seedhead formation can be prevented in the spring through the use of the plant growth regulator mefluidide (Embark). Using a base 50 degrees F (10°C), when 25 degree days have been accumulated, the mefluidide should be applied. By preventing seedhead formation, the annual bluegrass roots will continue to grow during the spring, developing a healthier plant that is better able to survive the summer stress period.

The Fine-Leaf Fescues

The fine-leaf fescues will make a turf suitable for unirrigated fairways in cool climates on sandy soils. It should be kept in mind that we are not talking about soft, lush green fairways. We are talking about fairways that will be a soft, lush green in the spring and fall, but during the summer months they will be hard and brown. A real artist with an irrigation system may be able to keep them slightly green in the summer, but hardly soft and lush. Attempts to keep them soft and lush will quickly turn a fine-leaf fescue fairway into a not-so-fine-leafed annual bluegrass fairway.

The fine-leaf fescues will compete better with annual bluegrass on lighter soils, especially sandy ones. On the really fine textured soils, it may be futile to attempt to keep out annual bluegrass.

Diseases

The fine-leaf fescues are susceptible to leaf spot, net-blotch, anthracnose, dollar spot, red thread, and rust.

Cultural Programs

Irrigation. If fine-leaf fescue fairways are to survive, they must receive as little water as possible. They should be watered only when severe drought stress is evident (when the grass plants show signs of wilt). They should be watered lightly, and not irrigated again until drought stress reappears.

Fertility. One of the nice things about fine-leaf fescues is that they will survive very well with the application of as little as 1 pound of actual nitrogen per 1000 square feet (.5 kg per 100 square meters) per season. If only 1 pound of nitrogen per 1000 square feet (.5 kg per 100 square meters) is to be applied, the ideal time to apply it is in late summer or early in the fall when the cool weather returns and the grass begins to grow vigorously again.

Cultural Practices. The only cultural practice that should be necessary on a fine-leaf fescue fairway is coring when and where compaction is a problem.

Perennial Ryegrass

Perennial ryegrasses are being used more and more as a fairway turf. They can tolerate the low height of cut, one-half inch or below, that is demanded by today's golfers. The new improved cultivars mow better than the old ones, although they still look terrible if mowed with dull equipment in the summer.

The perennial ryegrasses, at least the vigorous ones, have an advantage over creeping bentgrass where overseeding into established annual bluegrass is desired. Because of their ability to become established in existing mature turfs, the perennial ryegrasses can be interseeded annually, much as annual bluegrass does naturally. In fact, annual interseeding has become a part of perennial ryegrass cultural systems for golf course fairways.

The other advantage perennial ryegrass has over creeping bentgrass is that a highly selective herbicide is available that has post-emergence control of annual bluegrass in ryegrass turfs—Prograss. I know you have heard that Prograss can be used to control annual bluegrass in creeping bentgrass turfs as well. In my experience, the low rates required so as not to injure the bentgrass, will knock down the annual bluegrass during the winter so that it appears completely dead during spring greenup, only to recover and begin growth within a month. Prograss is really most effective in perennial ryegrass turfs where higher rates can be used.

The perennial ryegrasses have about the same warm weather adaption as creeping bentgrass and annual bluegrass, but appear to lack direct low temperature stress hardness of these species. Perennial ryegrass does not produce rhizomes or stolons, and therefore cannot fill in voids left by divots, mechanical injury, or disease. Annual bluegrass or weeds will probably fill in the voids if interseeding is not practiced.

Diseases

The perennial ryegrasses are the most susceptible grass species to *Pythium* blight and are equal to the tall fescue and St. Augustinegrass in susceptibility to brown patch (Figures 11.3 and 11.4). They are also susceptible to dollar spot, red thread, and rust; are very susceptible to *Typhula* blight; and require a snow mold fungicide treatment in areas where these diseases are likely to occur.

Figure 11.3. ***Pythium*** **blight with mycelium present on overseeded annual ryegrass.**

Figure 11.4. Brown blight (*Drechslera siccans*) thinning a new seeding of perennial ryegrass.

Maintenance Practices

Irrigation: The perennial ryegrasses have better drought tolerance than either creeping bentgrass or annual bluegrass. In spite of this, they have a limited root system in the warm weather, and respond best to a light daily irrigation regime.

Fertility: The perennial ryegrasses do not require a lot of nitrogen to survive or to maintain their color. They should receive 1 to 2 lb per 1000 square feet (.5 to 1 kg per 100 square meters) in spring and early summer, *none* in the midsummer to reduce the severity of *Pythium* blight and brown patch, and 1 to 2 lb per 1000 square feet (.5 to 1 kg per 100 square meters) in the fall to reduce the severity of red thread and rust.

Cultural Practices: Since the perennial ryegrasses have no rhizomes or stolons, they tend to produce little thatch. They can be core-aerified to relieve compaction. This is often done in conjunction with reseeding in the fall.

WARM-SEASON GRASSES: WINTER OVERSEEDING

There are two growing seasons for turfgrasses in the southern United States—the warm grass season and the overseeding season. During the overseeding season, a cool-season grass is overseeded into bermudagrass greens, tees, and fairways when they go dormant and discolor in the cool weather. The major diseases of cool-season grasses were discussed earlier—diseases that occur on overseeded cool-season grasses will be considered here.

Bermudagrasses are dormant during the fall and winter in most of the southern United States, but golf is played year-round in most of this region. Therefore, the greens, tees, and (sometimes) the fairways are overseeded. Overseeding of greens is done to provide a smooth putting surface, to provide green color, and to prevent excessive wear of the dormant bermudagrass. On the tees, active growth is required for recovery from injury and also for preventing excessive wear of the dormant bermudagrass. Fairway overseeding is expensive and usually is used sparingly. Greens are seeded heavily, tees less so, and fairways least of all.

Several grasses, including Italian ryegrass, perennial ryegrass, bentgrass, fine-leaf fescue, and rough bluegrass, are used to overseed dormant bermudagrass turfs. The most commonly used are the improved turf-type perennial ryegrasses, rough bluegrass, and the creeping bentgrasses. Overseeding can be started while the daytime temperatures are still in the 80s (27°C), and in some areas such temperatures can be accompanied by high humidity. Light midday irrigation is used to help establish the overseeded grass. All of these conditions favor the development of *Pythium* blight (cottony blight), and the disease can cause severe losses of the young, tender perennial ryegrass seedlings. When conditions are favorable for the development of *Pythium* blight, the overseeded greens should be protected with a *Pythium* blight fungicide.

Winter Overseeding Bermudagrass Greens

1. Verticut the greens lightly in at least two directions to open up the turf. Be sure to remove debris by mowing at the regularly maintained height of cut.
2. Use only the highest quality seed available, and utilize mixtures or blends of the better turfgrass species and cultivars. Calibrate the seeder to apply the seed at one-half of the regular seeding rate, then apply the seed uniformly in at least two different directions.
3. The seed can be worked into the bermudagrass turf with a drag mat made either of burlap or steel.
4. Topdress lightly with 1/4 to 1/8 inch (.6 to .3 cm) of sand.
5. Drag the top dressing lightly into the turf.
6. Water the greens carefully to prevent any movement of seed. Watering too heavily will float the seed to the surface and cause uneven seedling establishment.
7. Water the greens daily to keep the moisture levels constant during the critical germination period. A midday application usually is more useful, and greens watered at midday are not as likely to get *Pythium* blight or damping-off as are greens watered at night.
8. Play can continue, but be sure to move the cups daily to reduce any uneven wear.
9. Grass must be cut high (at least 1/4 inch [.6 cm]) until the seedlings become established. Be sure the mowers are sharp, and don't use the grasscatcher during the first few mowings. Try to mow when the grass is dry.
10. Do not fertilize with nitrogen at overseeding time because this will tend to encourage unwanted bermudagrass competition, as well as development of *Pythium* blight. Wait until the seedlings have become established before fertilizing, and water immediately after each fertilizer application.
11. Maintain a constant watch for the development of *Pythium* blight, or apply fungicides for *Pythium* blight management on a preventive basis.

WARM-SEASON GRASSES: GREENS

Creeping Bentgrass

The discussion of creeping bentgrasses in the cool-season turfgrass region (pp. 186–189) applies here also. Creeping bentgrasses are used on a permanent basis on golf courses in the cooler and middle regions of the warm climatic zone, and in the southwest semi-arid and arid desert areas where the low humidity and cooler nights allow creeping bentgrass to be grown on a permanent basis year-round. There has also been an attempt in recent years to extend the creeping bentgrass range into the warm humid areas of the southeastern United States. I think the jury is still out.

Two things that we have learned about growing creeping bentgrass in the warm humid zone are that water management and air movement are critical if the grass is to survive. Deep, frequent, or infrequent irrigation that reduces the oxygen in the soil will result in severe loss of the creeping bentgrass turf. In years of intense rainfall it will be difficult to maintain healthy creeping bentgrass greens.

The key to maintaining healthy creeping bentgrass in the southern U.S. is to use light daily irrigation. Most golf courses are equipped with mist irrigation heads around the greens. These should be used 1 to 4 times a day to cool the turf off. In more arid climates, some light day or nighttime waterings are needed to bring the total of syringe irrigation plus regular irrigation to between 50% and 60% of what is lost through evaporation from an open pan.

If a creeping bentgrass green is growing in a closed-in area with little or no air movement, it will not do very well. Many courses have installed fans to promote air circulation around the greens, and have had outstanding results.

Since most creeping bentgrass is grown on greens that have high sand content, they are susceptible to developing localized dry spots, or "hot spots," as they are known in the south. A good wetting agent program can help to prevent them from occurring. Even on a good wetting agent program, hand-watering the dry spots on a daily basis during the hot summer months will be necessary.

Suggested Cultivars

Penncross, Penncross, Penncross is suggested until someone demonstrates that another cultivar can tolerate the hot summer weather.

Diseases

Dollar spot, brown patch, and red leaf spot may require fungicide treatment. It should be kept in mind that brown patch and *Pythium* blight will be more severe in the warm climatic zone than in the cool-season turfgrass region. We have recently found severe problems with take-all patch disease on creeping bentgrass growing in the southeastern U.S.

Cultural Programs

Fertility. The maximum amount of nitrogen that is required in the cooler parts of the warm climatic zone is 4 to 5 pounds per 1000 square feet (2 to 2.4 kg per square meters) per season. In the warmer climates, where creeping bentgrass grows for 10 to 12 months a year, the maximum amount of nitrogen that normally is required is 5 to 7 pounds per 1000 square feet (2.4 to 3.4 kg per 100 square meters). The actual amount of nitrogen that is needed by a given turf depends upon the texture of the soil. I strongly encourage using nitrogen fertilizer on the basis of "pounds per growing month." However, the less nitrogen that can be used to maintain dense turf, the less severe the major disease problems will be. Adequate levels of phosphorus and potassium, as determined by soil tests, should be maintained.

Cultural Practices. These are the same as those for cool-season creeping bentgrasses. A well-drained, coarse-textured soil for the root zone is even more critical for creeping bentgrass in the warm climatic zone than it is in the cool-season turfgrass region.

Bermudagrass

The improved fine-leaf bermudagrasses are the preferred turfgrasses for golf course greens in the southern portion of the warm climatic zone, where it is simply too hot and humid to maintain creeping bentgrass.

Suggested Cultivars

All fine-leaf bermudagrasses are propagated vegetatively. The cultivars that are best for use on golf course greens are Tifgreen and Tifdwarf. Both have excellent low-temperature hardiness and minimal seedhead formation.

Diseases

Bermudagrass decline is the most devastating disease on bermudagrass. Bermudagrass is also susceptible to dollar spot, brown patch, *Helminthosporium* diseases, and *Pythium* blight. These diseases are not as great a problem as they are on creeping bentgrass, but bermudagrass is susceptible enough that a preventive fungicide program may have to be followed. Bermudagrasses grown in hot, arid regions have very few turfgrass diseases.

Cultural Programs

Mowing Height. The fine-leaf bermudagrasses should be mowed to a height of $^1/_8$ to $^1/_4$ inch (.3 to .6 cm).

Fertility. Bermudagrass greens, unlike creeping bentgrass greens, don't grow all winter in most areas of the warm climatic zone. As a consequence, the fertility schedule for bermudagrass is different from that for creeping bentgrass. Nitrogen fertility should be avoided during late summer and early fall to reduce the severity of spring dead spot where it occurs. Phosphorus and potassium should be maintained at adequate levels as determined by soil tests.

Cultural Practices. Bermudagrass greens require the same type of cultural system as do creeping bentgrass greens except they require more vertical mowing. This system includes coring to relieve compaction, vertical mowing to remove graininess and thatch, and topdressing to reduce thatch and provide a smooth putting surface. Irrigation should be applied as needed to prevent wilt. The quantity of water applied is not as critical as it is on creeping bentgrass greens where overwatering often leads to the invasion of annual bluegrass. Annual bluegrass behaves as a true winter annual in the warmer parts of the climatic zone, and it will not survive the summer heat.

WARM-SEASON GRASSES: TEES

Bermudagrass

Suggested Cultivars

The cultivars Tifway, Tifgreen, Santa Ana, Midiron, and Vamont are suggested for use on tees.

Diseases

See the discussion of diseases on bermudagrass greens.

Cultural Programs

Mowing Height. Bermudagrass tees should be mowed to a height of 1/4 to 1/2 inch (.6 to 1.3 cm).

Fertility. Bermudagrass tees need more nitrogen than bermudagrass greens to help fill in divots.

Cultural Practices. The cultural practices employed for bermudagrass greens should also be applied to bermudagrass tees. In addition, divot holes on the tees should be filled with soil.

Zoysiagrass

Zoysiagrass can be used on tees. However, it is slow to recuperate from damage, and probably should not be planted in areas where intensive use is anticipated. It possesses superior wear tolerance, but openings are slow to heal. On par-3 holes, where irons frequently are used, the zoysiagrass might be replaced by weedy grasses in the bare spots.

Suggested Cultivars

The cultivars FC13521, Emerald, Meyer, and El Toro are suggested for use on tees.

Diseases

Zoysiagrass is susceptible to zoysia patch, dollar spot, crown- and root-rot, rust, and brown patch.

Cultural Programs

Fertility. Zoysiagrass tees should receive 2 to 4 pounds of nitrogen per 1000 square feet (1 to 2 kg per 100 square meters) per season. Phosphorus and potassium should be maintained at adequate levels, as determined by soil tests.

Cultural Practices. Zoysiagrass tees should receive the same cultural practices as do bermudagrass greens, plus filling of divot holes with soil, and possibly even spot plugging.

WARM-SEASON GRASSES: FAIRWAYS

Bermudagrass

Diseases

Bermudagrass fairways are probably the finest fairways off which to play a golf shot. Bermudagrass is the preferred turfgrass species for golf course fairways in the short-dormancy areas of the warm-season grass belt and, if it were not for spring dead spot (SDS; see Figure 11.5), it would be the choice throughout the climatic zone. But because of spring dead spot many superintendents are looking at the feasibility of zoysiagrass as a fairway turf, at least in the cooler part of the warm-season grass belt and in the transition zone. All cultivars of bermudagrass are susceptible to SDS when they are grown in areas where bermudagrass goes dormant. The longer the dormant period, the more serious the SDS problem is. Applying DMI and benzimidazole fungicides in the midsummer through midfall has been shown to reduce the severity of SDS. The cost of such a program would be $10,000 to $20,000 per year. That is a small price to pay for bermudagrass fairways. It is also a lot cheaper than renovating fairways every 5 to 6 years.

Figure 11.5. Spring dead spot on a golf course tee (Photograph courtesy of Dr. Bill Small).

Bermudagrass also is susceptible to dollar spot, brown patch, *Helminthosporium* diseases, Bermudagrass decline, and spring dead spot. The severity of each disease depends upon the area in which bermudagrass is grown, but in general, disease pressure is greater in the warm humid climatic zones than in the semiarid and arid zones.

Suggested Cultivars

The cultivars Tifway, Vamont, Midiron, and Flora Tex are suitable for use on fairways.

Cultural Programs

Mowing Height. Bermudagrass fairways should be mowed to a height of 1/2 to 1 inch (1.3 to 2.5 cm).

Fertility. One pound of actual nitrogen per 1000 square feet (.5 kg per 100 square meters) per growing month is adequate. Last application should be made one month before the bermudagrass goes dormant to reduce the severity of spring dead spot.

Cultural Practices. The cultural program for bermudagrass fairways is the same as that for bermudagrass greens, except that topdressing should be omitted because it is too expensive. However, vertical mowing for thatch removal is necessary, especially for some of the more aggressive improved cultivars.

Zoysiagrass

Zoysiagrass may turn out to be the ideal grass species for golf course fairways in the transition zone and in cool regions of the warm-season grass belt. It has one major problem, a disease called zoysia patch. Many golf course superintendents who lost their bermudagrass fairways year after year to spring dead spot, tried zoysiagrass on their fairways, only to find big dead patches during the summer from zoysia patch. Once again, trying to solve a serious disease problem on one species by changing to another species only led to a different serious disease problem.

Zoysiagrass is a slow-growing species. It therefore is often sold to the homeowner on the basis of requiring less mowing than do other species. This same "good" quality on a home lawn can be an undesirable characteristic on a golf course fairway, because zoysiagrass injured mechanically or by disease or worn by traffic, will be slow to recover. This will leave voids in the turf that may be filled by weeds.

It will take many years to determine which species, bermudagrass or zoysiagrass, and which cultivars within those species are best adapted for golf course fairways in the cooler and middle regions of the warm-season grass belt and the transition zone. But whichever is chosen as "the" fairway grass, it will have diseases, and they will have to be managed!

Zoysiagrass fairways are established vegetatively, with either plugs or sprigs (stolons). Sprigs are preferable because they are less expensive to buy, and because they can be spread less expensively and more quickly than can plugs. The sprigs should not be completely buried because they will die unless some leaf material remains above the ground. The other important thing to remember when establishing zoysiagrass sprigs is to keep them moist for the first two weeks after establishment. If they become dry, they will die.

Diseases

Zoysiagrass is susceptible to crown-and-root-rot, rust, brown patch, and dollar spot. However, the most serious disease, and the one which is limiting the use of zoysiagrass, is zoysia patch. Hopefully, strategies to manage this disease will be developed in the not-too-distant future.

Cultural Programs

Irrigation. Zoysiagrass fairways should be irrigated lightly and frequently. They have very poor drought tolerance.

Mowing Height. Zoysiagrass fairways should be mowed to a height of 3/4 to 1 inch (1.9 to 2.3 cm).

Fertility. The application of 2 to 4 pounds of actual nitrogen per 1000 square feet (1 to 2 kg per 100 square meters) per season is suggested.

Cultural Practices. Zoysiagrass can develop a heavy thatch. It will require vertical mowing and coring, with return of the cores, to keep the level of thatch to a minimum.

Buffalograss

Buffalograss is the fine-leaf fescue of the warm-season grasses. Recently, some new cultivars with improved agronomic characteristics have been developed. It will be interesting to see what kind of disease resistance they have. Buffalograss persists on nonirrigated fairways in the more arid areas of the warm-season grassbelt. Buffalograss requires little water and nitrogen fertility; as a matter of fact, the quickest way to eliminate buffalograss is to irrigate it and fertilize with nitrogen. I doubt it will ever be accepted as a fairway grass on high maintenance golf courses; it may do well as a fairway grass on low maintenance golf courses. Its primary use may be as a grass for the roughs.

Suggested Cultivars

There are no buffalograss cultivars that can be recommended at the present time.

Maintenance Programs

Irrigation. Buffalograss requires little or no supplemental irrigation.

Mowing Height. Buffalograss fairways should be mowed to a height of 3/4 to 1 inch (1.9 to 2.5 cm).

Fertility. Buffalograss fairways should receive 1 to 2 pounds of nitrogen per 1000 square feet (.5 to 1 kg per 100 square meters) per season.

Cultural Practices. No cultural program has been developed for buffalograss.

Kikuyugrass

Kikuyugrass is the annual bluegrass of the warm-season grasses. Someone has decided it is undesirable, and since that time, great effort has been put into trying to eradicate it from golf course fairways, tees, and roughs with about as much success as people have had removing annual bluegrass from cool-season turfs. Unless some eradication program is initiated immediately after kikuyugrass is discovered on a golf course, it is futile to try to eradicate it due to extensive seeding capability at low mowing heights. Since there are no selective controls for removing kikuyugrass, an eradication program must consist of scouting a golf course in search of kikuyugrass, and treating with glyphosate. This is almost a daily process; I will guarantee you, it is not something you do once a year. It is a lifetime commitment, since there is no way to completely eliminate every last kikuyugrass plant from a golf course once it is present.

Once a golf course is 25% or more kikuyugrass, you might as well learn to live with it. I have seen people "kill" the kikuyugrass fairways with glyphosate, only to have the fairways return to kikuyugrass within three years. It is like removing a few buckets of water from the ocean; it might make you feel better, but you'll never notice the difference.

Cultural Program. There are no good data on managing kikuyugrass because, like annual bluegrass, it has been treated as a weed by researchers. What is given below is based on observation. Only when it is recognized for what it is, another acceptable turfgrass species, will we have the necessary research done to develop management programs.

Irrigation. Irrigate when the plants begin to show signs of wilt.

Mowing height. It can be mowed at a height of 1/2 inch (1.3 cm) for fairway use.

Fertility. Around one-half pound of actual nitrogen per 1000 square feet (.25 kg per 100 square meters) per growing month.

Diseases. Dollar spot and brown patch are listed as problems. I have observed a patch disease on kikuyugrass in South Africa.

CHAPTER 12

Disease Management Strategies for Home-Lawn Grasses

COOL-SEASON GRASSES

This chapter will deal with the advantages and disadvantages of the various lawn grasses, and the cultural, genetical, biological, and chemical practices necessary to maintain them. The adaptability of some turfgrass species to shade will be discussed also.

The Kentucky bluegrasses, perennial ryegrasses, tall fescues, and the fine-leaf fescues are the most important home lawn grasses in the cool-season turfgrass region. The bentgrasses (creeping and colonial) are too disease-prone and require too much maintenance to be desirable home-lawn turfgrasses.

The Kentucky Bluegrasses

Universities and private industry have devoted major effort to the improvement of the Kentucky bluegrasses, perennial ryegrasses, and turf-type tall fescues. As a result, there are many improved cultivars from which to choose. Unfortunately, there are many poor ones also. I will attempt here to separate the good from the bad and the ugly. It may be that some people will be unhappy with this chapter because their particular cultivar either was not mentioned or was not suggested for use, but in the long run, the facts about the various cultivars will become known anyway. Obviously, many new cultivars will become available after the publication of this book; you should therefore check with your local turf expert for the best cultivars for your area.

Environmental Adaptation

The Kentucky bluegrasses have wide adaptation. The diversity arises partially from the numerous cultivars, which give a broad genetic base from which to choose. Kentucky bluegrass can be grown in all parts of the cool-season turfgrass belt. It does best when grown in full sunlight. Certain cultivars can survive in open shade, but not in dense shade or where there is also tree root competition. The perennial ryegrasses are a better choice for shady areas, especially where there is tree root competition.

Cultivars

Kentucky bluegrass cultivars may be subdivided into two categories, the "improved types" and the "common types." In general, the "improved types" are those cultivars that have been bred or selected for resistance to melting-out. The "common types" are those cultivars that have little or no such resistance. The first "improved" cultivar or cultivar with resistance to melting-out was Merion, which was discovered by Joe Valentine on the Merion Golf Course in Pennsylvania. This discovery was really the beginning of the Kentucky bluegrass era for home lawns, and Merion has been widely used for many years. It started the sod industry on its way, because sod growers needed a grass that they could raise without chemical sprays.

The Kentucky bluegrasses, particularly Merion, were found to have other disease problems after the melting out-resistant cultivars became established. Diseases that once were considered to be of minor importance suddenly became major problems, and new diseases, like necrotic ring spot and summer patch (originally mistakenly called *Fusarium* blight) (Figure 12.1) began to appear. Stripe smut (and later flag smut), stem rust, and powdery mildew became problems because Merion is a single strain. The fact that Merion is a single strain with a narrow genetic base allowed these three pathogens (which are composed of many different genotypes, called races) to attack it. Those pathogenic races that could attack Merion eventually became the dominant ones in

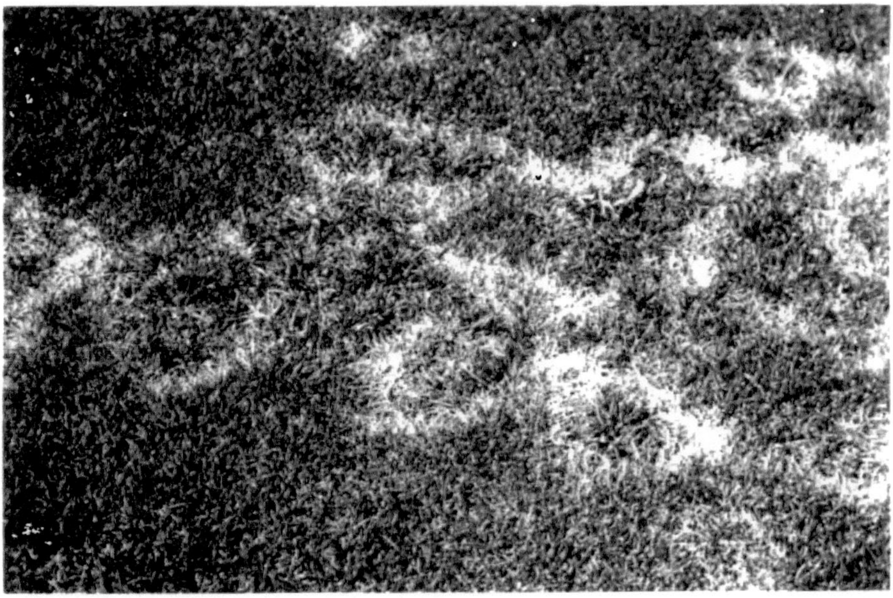

Figure 12.1. Kentucky bluegrass lawn infected with necrotic ring spot.

the population, and they parasitized Merion wherever it was grown. It became obvious that Merion had no horizontal resistance to these pathogens, and as a consequence, powdery mildew became a problem in the shade, stem rust became a problem under conditions of low fertility or where long periods of cool fall weather occurred, and stripe smut, like necrotic ring spot and summer patch, became a problem everywhere (see Figure 12.2).

Suggested Cultivars

Today there are four major diseases of Kentucky bluegrass: melting-out, necrotic ring spot, summer patch, and stripe smut. In order to have a lawn that can be sustained at a high level of quality, a cultivar that has resistance to all four diseases should be used whenever possible. Resistance to only one or two of them will result in an unsatisfactory lawn, because any one of the four major diseases is capable of destroying an entire lawn or making it so unattractive that you'll wish you had planted corn. At least then you would have something to harvest.

A blend of three or four of the Kentucky cultivars best adapted to your area should be used for best results. The use of blends over the

Figure 12.2. A stripe-smut-infected 'Merion' Kentucky bluegrass turf.

past 20 years or so has virtually eliminated stripe smut as a serious problem.

Do the common-type Kentucky bluegrasses have a place? The answer is very definitely yes! They will do much better than the improved types when grown under conditions of low maintenance, which usually means little or no supplemental irrigation or nitrogen. This is not to say that a common-type Kentucky bluegrass lawn will look as nice as the high-maintenance lawn. It will not, and it may eventually be taken over by weeds. It simply means that it will look better for a longer period of time under conditions of low maintenance than will the improved Kentucky bluegrass cultivars.

Cultural Programs

Fertility. The "improved" Kentucky bluegrass cultivars require 5 to 6 pounds of nitrogen per 1000 square feet (2.4 to 2.9 kg per 100 square meters) per season—approximately 1 pound per 1000 square feet (.5 kg per 100 square meters) per growing month. Maintaining adequate nitrogen levels is important if diseases like melting-out, necrotic ring spot, and summer patch are to be effectively managed. Research has clearly shown that chemical controls of these diseases will not be effective if adequate levels of nitrogen are not used. The slow release fertilizers have also been shown to be more effective in managing necrotic ring spot and summer patch than the quick release forms.

It is difficult to get the average homeowner to apply all the nitrogen a lawn needs. It is especially difficult to get them to apply nitrogen in the late summer when all they are thinking about is vacation, or in the fall when they are so sick of mowing the lawn that they are praying for frost. Late summer, fall, and dormant applications are important if the homeowner wants a minimal amount of disease problems. The dormant application is especially important. The timing for this application is after the last mowing, when the shoot growth mechanism has stopped. In other words, when you aren't going to mow anymore. This is the one time of year you can apply nitrogen and have it produce roots.

When nitrogen is applied during the growing season it normally produces shoot growth. When it is applied after shoot growth has stopped, the plant roots, which remain active for three to four weeks after shoot growth has ceased, continue to take up nitrogen and the plants continue to photosynthesize, producing carbohydrates that must go somewhere. Since the shoot growth mechanism is closed down, they go into the production of roots, rhizomes, and stolons.

It is difficult to get people to put on fall and dormant fertilizer applications. It is also resisted by lawn care companies because it doesn't fit their "normal" schedule, and they are often getting ready for the snow plowing season. It reminds me of the time I was giving a talk to a commercial lawn-applicators group. I suggested the need for six nitrogen applications, including a dormant application, much to the dismay of some of the people in attendance. One of them nervously stood up to try to "clarify" for the others what I had "really" said. He stated something to the effect of, "Dr. Vargas, you are talking only about fertility in relation to diseases," to which I answered with a smirk, "Yes, only in relation to diseases!" With a sigh of relief, he turned to the audience and said, "I hope you all understand that he is talking only about diseases." I said, "Yep, if you aren't worried about diseases there is no need to follow this schedule." Since my sarcastic remarks obviously went over his head, he sat down content that he had saved the day and that things would go on as they always had.

The issue is like having your car painted, and having someone advise you to apply rust proofing before you drive it through the salt of the first winter, only to have someone else stand up and say, "But you are talking only about preventing rust." What good is a nicely painted car that is rusted out? What good is an improperly fertilized lawn that is marred by disease? Yes, all I was talking about was preventing diseases—also known as "how you can have a lawn that you can be proud of."

Cultural Programs

Mowing Height. Mow to a height of 2 to 3 inches (5.1 to 7.6 cm) (note: not 1/2 to 1 inch [1.3 to 2.5 cm]). You are not going to putt, chip, or drive off your lawn. Since the ideal height of cut for the survival of Kentucky bluegrass is 2 to 3 inches (5.1 to 7.6 cm), mow it to that height. The lawn should be mowed frequently enough that no more than 1/3 of the shoot growth is removed at any one mowing.

Fertility. Kentucky bluegrass lawns require 4 to 6 pounds of nitrogen per 1000 square feet (2 to 3 kg per 100 square meters) per season.

Dethatching. Coring and dragging the cores to reincorporate the soil should be practiced where thatch is a problem.

Irrigation. Irrigation should be on a daily basis during the warm summer months. Fifty to sixty per cent of the water loss to evaporation should be replaced on a daily basis. This is usually between 0.1 and 0.4 of an inch (.25 and 1 cm) of water, the higher rates being applied in the more warm and arid climates and in drought years in cooler climates.

Kentucky Bluegrass Lawns with Existing Disease Problems

To minimize damage from the three major disease problems of Kentucky bluegrass turfs—melting-out, necrotic ring spot, and summer patch—approximately 1 pound of nitrogen per 1000 square feet (.5 kg per 100 square meters) per active growing month should be applied. Nitrogen in the early spring will decrease the severity of melting-out, nitrogen in the early summer and midsummer will help manage necrotic ring spot and summer patch.

Melting-out. The suggested treatment for melting-out is to apply a dormant nitrogen application in the fall and moderate applications in the spring. Light daily irrigation will also help reduce the severity of melting-out. In most seasons, nitrogen applications and daily irrigation will provide satisfactory management of melting-out. In years of severe disease incidence, fungicides like chlorothalonil, mancozeb, vinclozolin, iprodione, or anilazine can be applied in the spring until the warm weather of summer arrives. A fungicide program for the control of melting-out can be very time-consuming and costly, so "do it right the first time" and plant a melting-out resistant Kentucky bluegrass cultivar.

Necrotic ring spot. This is the one disease that we have a complete IPM program for managing. It can be managed biologically, culturally, genetically, and chemically. Furthermore, it can be managed biologically, culturally, and genetically without chemicals by simply following the suggestions in this section.

It was noted in Chapter 4 that the symptoms of necrotic ring spot are most severe during the summer when the infected plants are under drought stress. Light, daily irrigation during the day will help to suppress the development of symptoms. Deep, infrequent irrigation is of little use because the infected plants have short roots that cannot use irrigation water when it is 3 to 8 inches (7.6 to 20 cm) deep in the soil. Kentucky bluegrass turf infected with necrotic ring spot needs no more than 20 minutes of irrigation, but it needs it daily to keep the root zone of the infected plants moist so they can take up water. The ideal time for irrigation is the warmest part of the day. This not only supplies irrigation to the infected root system, it cools the turf and allows it to survive the warm weather stress. Daily irrigation, by keeping the thatch and soil moist, allows for the maintenance of high populations of beneficial bacteria that produce substances (antibiotics) that are toxic to *L. korrae*. A word of caution: a 10,000-square-foot (930 square meters) lawn infected with necrotic ring spot cannot be irrigated adequately with one dinky little sprinkler!

Nitrogen fertility level is also important in managing the disease. Kentucky bluegrass should receive between 4 to 6 lb of nitrogen per thousand square feet (2 to 3 kg per 100 square meters) per season. In general, a pound of nitrogen (.23 kg) per growing month, plus a dormant application is necessary. The slow release fertilizers have been shown to be more effective than the quick release.

The organic fertilizers, like Lawn Restore, have been shown to cause an increase in the number of beneficial bacteria in the thatch and soil. Many of these bacteria have been shown to produce compounds (antibiotics) that are toxic to *L. korrae*. Adequate nitrogen fertility, particularly with products like Lawn Restore, combined with light daily irrigation can manage necrotic ring spot without the use of fungicides.

Chemical management can be obtained on a curative basis with the benzimidazole systemic fungicides (benomyl and thiophanate methyl), or on a preventive basis with the DMI fungicides (fenarimol and propi-

Figure 12.3. Renovation and reseeding, with a commercial seeder, of a lawn infected with necrotic ring spot.

conazole). Management with iprodione and triadimefon has also been reported, although instances of their not managing the disease are also in the literature. Hopefully, someday we will figure out the reasons for the discrepancies.

To manage necrotic ring spot with benomyl or thiophanate methyl, the infected turf should be irrigated the night before the fungicide is applied, and these systemic fungicides will have to be watered in before they have a chance to dry on the foliage. Necrotic ring spot is a crown-and-root-rot problem, and the fungicide should be applied to the affected plant parts. The benzimidazole systemic fungicides will only be translocated upward; they are not translocated downward. This treatment is very expensive, and because of the expense, many people assume that one treatment will cure their problem forever. It won't, and retreatment will be necessary every year unless the cultural and biological management practices mentioned above are practiced.

Summer Patch. See Chapter 4 for information on how to manage this disease. The same biological and chemical management practices that are effective in managing this disease on annual bluegrass will also manage this disease on Kentucky bluegrass. The only difference is in the amount of nitrogen applied under cultural controls for the different species. Annual bluegrass should only have 1/2 lb of actual nitrogen per 1000 square feet (.25 kg per 100 square meters) per month in June, July, and August, whereas Kentucky bluegrass should have 1 lb of actual nitrogen per 1000 square feet (.5 kg per 100 square meters) per month in June, July, and August.

Stripe Smut. Having stripe smut in your lawn is a little like having a serious case of diabetes. You can live with it, provided you maintain a proper diet and take good care of yourself. A turfgrass plant infected with stripe smut is infected for life. All plants arising from the infected mother plant via rhizomes will also be infected. Stripe smut is a systemic disease that can remain dormant in the crown of the plant during the warm weather of summer or the cold weather of fall. Or it can spread up into the leaves, eventually rupturing the epidermis and releasing many black spores that may attack other plants. Whether the spores are visible or not, the plant is always in a weakened condition, and the first stress to which it is subjected may kill it. The most common stress is summer drought. Many people who take good care of their lawns and water them religiously while they are home, go away on vacation and forget them. If a healthy lawn of Kentucky bluegrass is not irrigated, it simply will go dormant and revive when water is applied again. However, if a lawn is infected with stripe smut and the same thing is allowed to happen, instead of going dormant it will go dead! It is, therefore, important that a lawn infected with stripe smut not be allowed to dry out.

Stripe smut can be managed (really, arrested) with the DMI fungicides if they are applied bimonthly during the growing season. This will need to be done every year as long as you have the lawn. As you have probably already figured out, this can be very expensive and time-consuming. The first choice should be the renovation of the lawn, and reestablishment with a blend of smut-resistant cultivars. The smut fungus produces many races which tend to be cultivar-specific. If you have four cultivars in a blend, the chances of the specific race of *U. striiformis* finding a cultivar it can attack are one in four. If it is successful in finding a cultivar it can infect, the chance of the new inoculum finding a cultivar are still only 1 in 4. Even if it found every plant of that cultivar to infect in the blend, which is highly unlikely, 3/4 of the cultivars would be left in the blend, and they could easily fill in the voids.

To show you how successful this approach has been, stripe smut was one of the worst diseases in the midwest in the late 60s and early 70s, when monostands of Merion, Windsor, and Fylking were grown. With the widespread use of Kentucky bluegrass blends since that time stripe smut has become an almost nonexistent disease.

Rusts. In most areas, the rusts can be managed adequately with nitrogen fertility. Rust is usually not a problem if enough nitrogen is applied so that the turfgrass is mowed at least once a week. However, cold temperatures, and not lack of nitrogen, is responsible for the slow growth of turfgrass in some coastal and northern areas. In these areas the DMI fungicides triadimefon, fenarimol, and propiconazole can be used in the fall of the year. However, in an era of environmental concerns, since leaf and stem rust are minor problems, you may wish to just tolerate these diseases when they occur late in the season.

Red Thread. Red thread, like rust, can be controlled in most areas with adequate levels of nitrogen. A fungicide must be used in far-northern and coastal areas where the slow shoot growth is due to cool weather, not to inadequate amounts of nitrogen. The benzimidazoles, dicarboximides, or DMI fungicides will manage the disease. Again, since it is such a minor disease, on well fertilized Kentucky bluegrass turfs, fungicides should probably not be used.

Powdery Mildew. Powdery mildew is a problem primarily in the shade. It will occur in open areas during falls that have many days of overcast weather. It is not a serious problem when it occurs in open areas, but can be a problem in the shade. It is impractical to try to manage powdery mildew in the shade. It is too expensive and time-consuming. There may someday be Kentucky bluegrass cultivars that have both shade adaptation and resistance to powdery mildew and melting-out. As things stand now, Glade is shade-tolerant and has resistance to powdery mildew, but it lacks resistance to melting-out; Nuggett is resistant to melting-out but it lacks resistance to powdery mildew, and is susceptible to necrotic ring spot. Bensun (Warren's A-34) appears to have resistance to both melting-out and powdery mildew, but is susceptible to necrotic ring spot. Bensun Kentucky bluegrass may be the best choice for planting in moist, dense shade in the cooler areas of the cool-season grass belt. The improved cultivars of *Poa trivialis* may be the best choices in the middle and southern regions. The fine-leaf fescues will do well in open shade, but leaf spot and powdery mildew can destroy them in dense, moist shade. Some of the new, improved perennial ryegrasses may offer a better alternative to all of the above, especially where tree root competition is also a problem.

Fairy Rings. Fairy rings are not a problem of grass as much as they are a problem of the soil. The fungi that cause fairy ring grow on organic matter in the soil or thatch. They release nitrogen or other substances as they grow, and this in turn causes the zone of stimulation (darker, taller grass) to form outside the main body of the fungus. The turf is usually yellow, wilting, or dead where the main body of the fungus is growing in the thatch (see Figure 12.4). The yellow color develops because the fungus interferes with water and nutrient uptake. Wilting and death of the grass occurs when the body (mycelium) of the fungus prevents moisture from reaching the root system. The only way to get rid of the fungus is to physically remove the soil that contains it. This can be accomplished by digging out the soil and sod from an area that is one foot deep and extends one foot beyond the fairy ring. The excised soil then is replaced by healthy soil and sod. If the fairy ring covers a large area, you may wish to remove the sod, till the area, and have it fumigated with methyl bromide. Methyl bromide is a deadly gas, and should be applied only by licensed personnel. There has been a lot written about drilling holes, and dumping this or that fungicide down the holes to kill fairy rings. It won't work.

Figure 12.4. Heavy infestation of nitrogen-induced *Tricholoma* fairy ring in a sod field.

Fine-Leaf Fescues

There are four species of fine-leaf fescue that commonly are referred to as red fescue. These are *Festuca rubra* L. subspecies *commutata* Gaud. (chewings fescue), *Festuca rubra* L. subspecies *rubra* (strong creeping red fescue), *Festuca rubra* L. subspecies *trichophylla* (slender red fescue), and *Festuca longifolia* Thuill. (hard fescue). Because of their susceptibility to leaf spot and other diseases, the fine-leaf fescues are used mostly on low maintenance home lawns, either alone or in mixtures with the common-type Kentucky bluegrasses. The fescues are especially well adapted to sandy soil, and they can survive nicely on one to three pounds of nitrogen per 1000 square feet (.5 to 1.5 kg per 100 square meters) per season. The fine-leaf fescues do well also in areas of open shade; in fact, they do better under conditions of full sunlight than they do in the shade until they are grown in the southern part of the cool-season climatic region, where open shade is the only environment in which they will survive.

Attempts have been made to select and improve the fine-leaf fescues to make them more desirable turfgrasses. Most of the selecting has been done on the basis of such agronomic characteristics as color, texture, and creeping habit, and very little selection has been done on the basis of resistance to disease. Until there are some cultivars developed with some disease resistance, the fine-leaf fescues will remain a turf for low-maintenance areas.

Leaf spot is assumed by most researchers to be the most important disease of the fine-leaf fescues in most areas today, and it could be the limiting factor in the use of these turfgrasses in home lawns. It may be, though, that everyone is looking for resistance to leaf spot and forgetting about the other known diseases of the fine-leaf fescues—diseases such as red thread, rust, *Microdochium* patch, anthracnose, and dollar spot. These are diseases mainly of slow-growing or low-maintenance turf, which is primarily where the fine-leaf fescues are grown. It is necessary that a lawn grass have resistance to leaf spot, but resistance to these other diseases will also have to be found if the fine-leaf fescues are to be more widely used.

If cultivars that are resistant to the major diseases are found, the fine-leaf fescues could be the home-lawn turfgrasses of the future. They will become more desirable as resources, like water and fertilizer, become less available and more expensive.

Maintenance Programs

Mowing Height. The fine-leaf fescues should be mowed to a height of 2 to 3 inches (5 to 7.5 cm).

Fertility. The fine-leaf fescues require no more than 1 to 3 pounds of nitrogen per 1000 square feet (.5 to 1.5 kg per 100 square meters) per season, and the less nitrogen applied the better. Other nutrients (mainly phosphorus and potassium) should be added as needed, as established by soil tests.

Irrigation. Irrigate lightly and daily; the fine-leaf fescues have excellent drought tolerance. The two quickest ways to destroy a fine-leaf fescue are overwatering and overfertilizing.

Suggested Cultivars

See Table 10.2 for information on fine-leaf fescue cultivars.

Perennial Ryegrasses

The new perennial ryegrasses have improved mowability. That is, they mow better than the old "common" type, which mows like baling wire, but they still don't mow like the Kentucky bluegrasses and fine-leaf fescues. They still look bad in midsummer if they are mowed with a reel mower, but they look terrible if mowed with a dull rotary mower. Perhaps the shabbiness could be tolerated if the improved perennial ryegrasses had excellent disease resistances and good turf density, but they don't. It is another case of turfgrass being developed for agronomic characteristics with little regard for susceptibility to disease. The perennial ryegrasses may someday be an acceptable turfgrass species for home lawns, but that someday isn't here yet. They are best suited for golf courses where mowers can be kept sharp and where fungicide sprays can routinely be made for dollar spot, brown patch, *Pythium* blight, rust, red thread, and *Typhula* blight. For more information, see Chapter 11 under golf course turfs.

Creeping Bentgrass

If you don't have a creeping bentgrass lawn, don't plant one unless you have enough money to hire a professional landscaper to take care of it, or enough expertise, time and money to take care of it yourself. If you want a nice-looking bentgrass lawn you must treat it like a golf course green. This means that in addition to mowing, irrigating, and fertilizing, you must also topdress and vertically mow. If creeping bentgrass is not topdressed and vertically mowed, it will become thatchy and puffy and it will scalp in the summer. It will be susceptible to leaf spot, dollar spot, brown patch, *Pythium* blight, and the snow molds. For a maintenance schedule and disease-control program, see the discussion of golf course greens in Chapter 11.

Tall Fescue

The new improved turf-type tall fescues are becoming more widely grown than their predecessors, which were primarily limited to general turf areas. They have become the turfgrass of choice in the Kentucky bluegrass-bermudagrass transition zone, where the bermudagrass often died of winter kill or spring dead spot, and the Kentucky bluegrass of summer patch and heat stress. This is not to say the turf-type tall fescues are not without disease problems. The limiting factor in the successful growth of tall fescue is brown patch. The tall fescues, along with the perennial ryegrasses, are the most susceptible species to brown patch. Brown patch is most severe under conditions of high humidity and high nighttime temperatures—the exact conditions you have in the area the tall fescues are adapted to. Hopefully, breeding will eventually solve the problem. In the meantime, the tall fescues, in spite of their susceptibility to brown patch, are still the turfgrass of choice for the Kentucky bluegrass-bermudagrass transition zone.

The turf-type tall fescues are not limited to the transition zone. They have a deep root system, even during the warm summer months, which makes them ideally suited for turf areas where little or no supplemental irrigation is used. This species is well suited for general turf areas like parks or roadsides. This is not to say that they can't be used for home-lawn turfs as well.

One other advantage of the tall fescues is that, like perennial ryegrass, some cultivars contain endophytes that protect them from foliar insects.

Maintenance Programs

Mowing height. Tall fescues should be mowed between 2 to 3 inches (5 to 7.5 cm).

Fertility. Tall fescue requires the application of 1 to 3 pounds of nitrogen per 1000 square feet (.5 to 1.5 kg per 100 square meters) per year.

Phosphorus and potassium should be added as the need for them is established by soil tests. All nitrogen fertilizer should be eliminated during the months of the year when brown patch normally occurs.

Irrigation. Irrigate as little as possible. Tall fescue is extremely drought-tolerant.

Diseases

Tall fescue is very susceptible to brown patch, which has limited its use. It is also susceptible to net-blotch and leaf spot.

Poa trivialis Rough Bluegrass

Normally considered to be a weed when it occurs in high-quality lawns, rough bluegrass does have a place as a turfgrass in dense, moist shade, especially in the middle and southern regions of the cool climatic zone. Very few other turfgrass species or cultivars will survive in heavily shaded areas. Rough bluegrass has been reported to be susceptible to leaf spot, rust, stripe smut, brown patch, *Microdochium* patch, and *Typhula* blight. The real truth of the matter is that it has not been grown widely enough, or in high-quality turfgrass areas, so its important disease problems remain unknown.

Cultural Programs

Mowing height. Mow at 2 to 3 inches (5 to 7.5 cm).

Fertility. Apply 1 to 3 pounds of nitrogen per 1000 square feet (.5 to 1.5 kg per 100 square meters) per season. The less nitrogen applied, the better. Potassium and phosphorus should be added when the need for them is established by soil tests.

Irrigation. Irrigate it! It grows best when it is kept moist.

Zoysiagrass

Zoysiagrass is fine as a warm-season turfgrass, but it turns brown at the first frost and remains so until late spring or early summer. If you want a lawn that is green for 2 to 5 months, then zoysiagrass is what you are looking for, but if you want a lawn that is green for the entire growing season, then zoysiagrass isn't for you. Zoysiagrass is discussed later in this chapter as a desirable turfgrass species for warm-season lawns.

WARM-SEASON GRASSES

Dealing with home-lawn diseases of warm-season grasses is not as easy as dealing with home-lawn diseases of cool-season grasses. First, most cool-season lawns are composed of Kentucky bluegrass. Secondly, more information is available on the disease resistance and susceptibility of cool-season cultivars. Warm-season home lawns are composed of many different species; for example, bermudagrass, centipedegrass, bahiagrass, St. Augustinegrass, and zoysiagrass. It is too early to tell, from the point of view either of plant pathology or agronomy, which grass is the superior one or which cultivar of the superior species is to be preferred. It may even turn out that one species is best for the hot, humid regions and another is best for the hot, arid regions.

Bermudagrass

The limiting factor in the use of bermudagrass on home lawns in the cooler regions of warm season grasses (or anywhere else, for that matter) is a devastating disease called spring dead spot (SDS; see Figure 12.5). It is a problem on bermudagrass everywhere except the warmer regions of the warm climatic zone, where bermudagrass has a short (1

Figure 12.5. Spring dead spot (SDS) on bermudagrass lawn (Photograph courtesy of Dr. R.V. Sturgeon).

to 2 month) dormant period, or no dormant period at all. The longer the dormant period, the more serious SDS becomes.

The bermudagrasses, especially the more widely grown cultivars, are notorious thatch producers and usually require some kind of yearly mechanical dethatching. Since most homeowners don't own the necessary equipment, either it must be rented or the homeowner must pay someone to do the job. This is an expense that a homeowner should take into account before establishing a bermudagrass lawn. Since there are seed types and low nitrogen stress-tolerant cultivars that are not thatch producers, they should be chosen over other cultivars for home lawn use. Check with your local turfgrass experts for information about cultivars adapted for your area.

Spring Dead Spot

Several fungi have been implicated in causing spring dead spot. They all have one thing in common: they attack the roots and the crowns of the plants from the midsummer to early fall. The weakened plants die when subjected to the cold weather of winter. When the rest of the turf breaks dormancy and begins to grow in the spring, those areas that were killed by the fungi become evident. Late summer and early fall applications of fungicides like benomyl, thiophante-methyl, fenarimol, and propiconazole have been shown to manage SDS. The effectiveness of any one of these fungicides many vary, depending on the fungus responsible for the disease in your area. You should check with the turfgrass experts in your area for the best treatment.

Bermudagrass Decline

A relatively new disease, bermudagrass decline is caused by *Gaeumannomyces graminis*, which thins and kills bermudagrass during the growing season. These patches may be what later becomes spring dead spot during the dormant season. Unfortunately, at this time we know little about fungicide treatments.

Other Diseases

Brown patch, leaf spot, and dollar spot are serious problems on bermudagrass in the humid regions of the warm climatic zone. To prevent severe loss of grass, use one of the fungicides suggested in Table 6.8.

St. Augustinegrass

St. Augustinegrass decline (SAD; see Figure 12.6) attacks St. Augustinegrass in Texas, New Mexico, and Louisiana. It is a viral disease, and there is no known chemical control for it once a plant has become infected. SAD is transmitted mechanically by mowing. Once it occurs in a small area of a turf, it is usually spread very rapidly throughout the entire turf area by the mowers. However, it may take 3 to 5 years before the turf is badly thinned, if it is maintained at a high cultural intensity.

Gray leaf spot, which is caused by *Pyricularia grisea*, also is a serious problem on St. Augustinegrass, especially on newly sprigged areas or where the grass is succulent and growing rapidly because of nitrogen

Figure 12.6. Foliar symptoms of St. Augustine decline virus (SAD) on a St. Augustine-grass blade (Photograph courtesy of Dr. Bobby G. Joyner).

fertilization. The disease is most serious at temperatures between 70° and 80°F (21° to 26°C) under conditions of high humidity. Irrigation during the daytime, allowing the foliage to dry more rapidly, and avoidance of high nitrogen fertility during the warm months will aid in reducing the severity of the disease. Mancozeb, chlorothalonil, or anilazine can be used on a 7- to 10-day schedule to manage the disease.

Brown patch (Figure 12.7), which is caused by *Rhizoctonia solani*, is the most serious problem on St. Augustinegrass during the spring and early fall when the temperatures are between 75° and 90°F (23.8° and 32.2°C), and humidity is high. The disease is most severe during rainy periods when nocturnal temperatures remain high, and on poorly drained soils. High nitrogen fertility and shade increase the severity of brown patch, as they do with gray leaf spot. Brown patch can be controlled by applying any of many fungicides (e.g., chlorothalonil, iprodione, anilazine, or mancozeb) at 7- to 10-day intervals.

St. Augustinegrass is one of the preferred species for home lawns and general turf in the southern portions of the warm, humid climatic zone, especially along the Gulf Coast. Floratam has excellent resistance to SAD and chinch bugs, either one of which will destroy St. Augustinegrass, but it lacks good agronomic characteristics. Also, it does well only in the warmer areas of the St. Augustinegrass growing region,

Figure 12.7. Brown patch on St. Augustinegrass.

because it lacks direct low-temperature stress hardiness. Floralawn, Raleigh, and Serville are new cultivars with resistance to SAD. Avoidance of excess nitrogen fertility in early fall helps reduce the severity of gray leaf spot and brown patch.

Zoysiagrass

Zoysiagrass is the most disease-resistant of all the warm-season turfgrasses that are in widespread use. It is the most cold-temperature-tolerant of all the warm-season grasses that are being grown from the midregions of the cool climatic zone to the southern regions of the warm-season grass belt. While zoysiagrass will grow in the cool climatic zone, it is not suggested for use there because it discolors (turns brown) with the coming of cool weather in the late summer or early fall, and doesn't green-up until the warm weather of summer comes again. However, it is an excellent turfgrass for lawns in the warm-season grass regions where rust is not a severe problem.

Zoysia Patch

As you might expect, zoysiagrass has its patch disease. It was only a matter of the species being more widely grown before a devastating disease would be found. The disease is caused by *Gaeumannomyces* spp. and\or *Rhizoctonia* spp., and can affect large patches up to several feet in diameter during the warm weather. Early application of fungicides like fenarimol, triadimefon, and propiconazole have shown some success in managing the disease.

Rust and Brown Patch

Zoysiagrass is susceptible to brown patch and rust. Before it became more widely grown, we used to think that these were the only two economically important diseases of zoysiagrass. Brown patch can be a severe problem during periods of hot, humid weather when the temperatures remain between 75° and 90°F (23.8° and 32.2°C). The avoidance of excess nitrogen just before and during this period will aid in reducing the severity of disease. Zoysiagrass is a slow-growing turfgrass, which is fine, especially if you are not fanatic about mowing, but the slow shoot growth rate makes rust a more severe problem on zoysiagrass than it is on most other grass species. The slow growth of zoysiagrass precludes, in most cases, the use of nitrogen fertility to manage rust. A fungicide program using chlorothalonil or mancozeb on a 7- to

10-day basis, or triadimefon, fenarimol, or propiconazole on a 3- to 4-week basis will be necessary.

Suggested Cultivars

Meyer, Emerald, and El Toro are improved zoysiagrass cultivars.

Centipedegrass and Bahiagrass

Centipedegrass and bahiagrass are used on low-maintenance turf areas; hence, little is known about their disease problems, nor is much concern given to them. This situation may change, and more information may be forthcoming, especially about centipedegrass.

Appendix

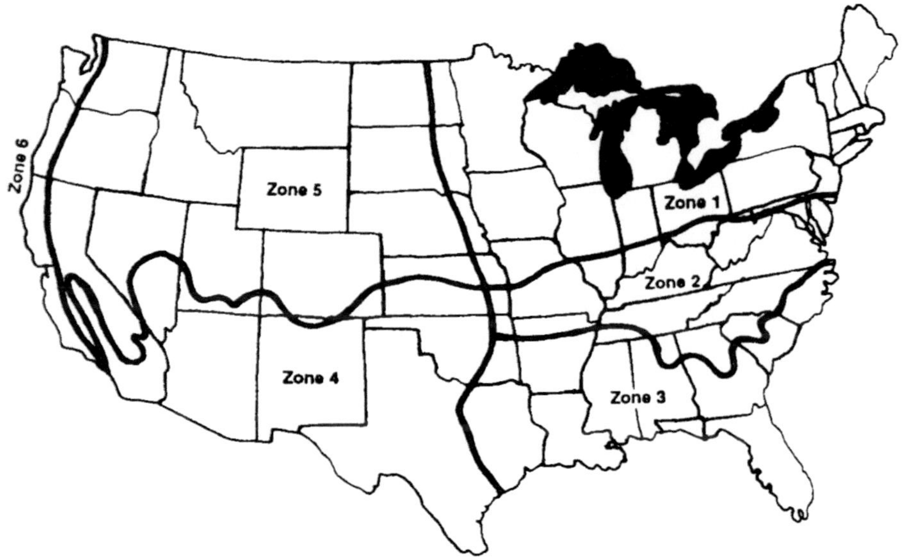

Figure A.1. The six climatic zones referred to in the nitrogen and fungicide schedules given in Chapters 11 and 12

Zone 1–Cool humid climatic zone
Zone 2–Bermudagrass–Kentucky bluegrass transition zone
Zone 3–Warm humid climatic zone
Zone 4–Warm arid climatic zone
Zone 5–Cool arid climatic zone
Zone 6–Cool oceanic climatic zone

Table A.1. Guide for the Control of Turfgrass Diseases

Diseases	Susceptible Turfgrass	Cultural	Chemical
Algae	All turfgrass	Reduce shade. Avoid excessive watering. Improve soil drainage.	Mancozeb, chlorothalonil
Anthracnose (*Colletotrichum graminicola*)	Annual bluegrasses, creeping bentgrasses, tall fescues, fine-leaf fescue	Fertilize and irrigate to maintain vigor.	Triadimefon, fenarimol, thiophanate methyl, propiconazole, benomyl
Brown patch (*Rhizoctonia solani*)	Creeping bentgrasses, Kentucky bluegrasses, tall fescues, perennial ryegrasses	Reduce nitrogen levels. Increase air circulation by removing surrounding vegetation.	Anilazine, chlorothalonil, mancozeb, iprodione, flutolanil
Dollar spot (*Sclerotinia homoeocarpa*)	Creeping bentgrasses, annual bluegrasses, perennial ryegrasses, fine-leaf fescues, colonial bentgrass, Kentucky bluegrass, bermudagrass, zoysiagrass	Maintain adequate levels of nitrogen. Remove dew.	Chlorothalonil, propiconazole, anilazine, fenarimol, iprodione, triadimefon, vinclozolin
Fairy rings (Basidiomycetes—soil fungi)	All turfgrasses	Remove infested sod and soil; replace with clean soil, and reseed or sod. Improve water penetration with aerification or wetting agents.	Methyl bromide or formaldehyde fumigation will eradicate fungus but also kill turf. (*Must be licensed to apply.*)
Gray leaf spot (*Pyricularia grisea*)	St. Augustinegrass, perennial ryegrass	Reduce nitrogen. Daytime irrigation.	Chlorothalonil, mancozeb, anilazine
Leaf spot (*Bipolaris sorokiniana*)	Kentucky bluegrasses, perennial ryegrasses, fine-leaf fescue, creeping bentgrasses	Remove clippings. Raise cutting height. Provide adequate levels of nitrogen.	Iprodione, chlorothalonil, anilazine, mancozeb, vinclozolin
Melting-out (*Drechslera poae*)	Kentucky bluegrass	Same as above	Same as above

Table A.1. Continued

Diseases	Susceptible Turfgrass	Cultural	Chemical
Microdochium patch (pink snow mold) (*Microdochium nivale*)	Annual bluegrasses, creeping bentgrass, fine-leaf fescues, perennial ryegrasses, Kentucky bluegrass	Avoid fall nitrogen applications. Rake leaves and cut short. Control drifting snow; may occur without snow cover.	Triadimefon, benomyl, fenarimol, iprodione, mancozeb, mercury chloride, pentachloronitrobenzene (PCNB), thiram, thiophanate methyl, vinclozolin
Necrotic ring spot (*Leptosphaeria korrae*)	Kentucky bluegrass, annual bluegrass	Provide adequate nitrogen. Provide light daily irrigation. Fertilize with Lawn Restore.	Fenarimol, iprodione, Propiconazole, thiophanate methyl, benomyl
Pink patch (*Limonomyces roseipellis, L. culmeginus*)	Red fescue, ryegrasses (perennial)	Provide adequate levels of nitrogen.	Triadimefon, propiconazole, fenarimol
Powdery mildew (*Erysiphe graminis*)	Kentucky bluegrasses, fine-leaf fescues	Reduce shade. Increase circulation by removing surrounding vegetation.	Triadimefon, fenarimol, propiconazole
Pythium blight (*Pythium aphanidermatum*)	Annual bluegrasses, perennial ryegrasses, fine-leaf fescues	Improve soil drainage. Increase circulation by removing surrounding vegetation.	Chloroneb, ethazole, propamocarb, metalaxyl, fosetyl-Al
Red leaf spot (*Drechslera erythrospila*)	Creeping bentgrasses	Fertilize to maintain vigor.	Iprodione, anilazine, chlorothalonil, vinclozolin
Red thread (*Laetisaria fuciformis*)	Fine-leaf fescues, perennial ryegrasses, Kentucky bluegrasses, annual bluegrass, creeping bentgrass	Remove clippings. Fertilize to maintain vigor.	Iprodione, anilazine, chlorothalonil, triadimefon, fenarimol, propiconazole
Rust (*Puccinia* spp.)	Kentucky bluegrasses, perennial ryegrasses, zoysiagrass, bermudagrass	Provide adequate levels of nitrogen. Mow frequently.	Mancozeb, chlorothalonil, triadimefon, fenarimol, propiconazole
Slime molds (*Myxomycete* spp.)	All turfgrasses	Remove mechanically by mowing or raking.	

Table A.1. Continued

Diseases	Susceptible Turfgrass	Cultural	Chemical
Spring dead spot	Bermudagrass	Avoid late season nitrogen.	Fenarimol, propiconazole, benomyl, thiophanate methyl
Stripe smut (*Ustilago striiformis*)	Kentucky bluegrasses, creeping bentgrasses	Reduce nitrogen. Irrigate to prevent dormancy.	Triadimefon, fenarimol, propiconazole
Summer patch (*Magnaporthe poae*)	Kentucky bluegrasses, annual bluegrass	Apply light, frequent watering during dry periods to reduce heat stress. Do not water heavily or deeply. Provide adequate nitrogen.	1. Fenarimol, triadimefon, propiconazole before symptoms are evident; 2. benomyl, thiophanate methyl, after symptoms are evident.
Take-all patch (*Gaeumannomyces graminis*)	Bentgrasses creeping, colonial bentgrass	Avoid drought stress.	Fenarimol, propiconazole
Typhula blight (*Typhula incarnata*, *Typhula ishikariensis*)	Creeping bentgrasses, annual bluegrasses, fine-leaf fescues, perennial ryegrasses	Avoid early fall nitrogen applications. Rake leaves and cut short. Control drifting snow.	Pentachloronitrobenzene (PCNB) (Proturf FFII), mercury chlorides (Calo-chlor, Calo-gran)[a], phenyl mercury (PMAS), chlorothalonil-iprodione.
Yellow tuft (*Sclerophthora macrospora*)	Annual bluegrass, Kentucky bluegrass, creeping bentgrass	Avoid standing water.	Metalaxyl

[a]Restricted use pesticides.

Bibliography

1. Alexander, P.M. 1969. Anthracnose: Serious disease problem. *USGA Green Section Record* 7(5):8–9.
2. Bean, G.A., and R.D. Wilcoxson. 1964. Pathogenicity of three species of *Helminthosporium* on roots of bluegrass. *Phytopathology* 54:1084–1985.
3. Beard, J.B. 1966. Fungicide and fertilizer applications as they affect *Typhula* snow mold control on turf. *Quart. Bull. Mich. Agric. Exp. Sta.* 49:221–228.
4. Beard, J.B. 1973. *Turfgrass: Science and Culture.* Prentice-Hall, Englewood Cliffs, N.J., p. 658.
5. Beard, J.B. 1976. Turfgrass Wear. *Mich. Turfgrass Conf. Proc.* 5:68–72.
6. Beard, J.B., P.E. Rieke, A.J. Turgeon, and J.M. Vargas, Jr. 1978. Annual Bluegrass *(Poa annua* L.): Description, adaptation, culture and control. *Mich. State Univ. Agric. Exp. Sta. Res. Rep. No.* 352, 31 pp.
7. Beard, J.B., J.M. Vargas, Jr., and P.E. Rieke. 1973. Influence of the nitrogen fertility level on *Tricholoma* fairy ring development in Merion Kentucky bluegrass *(Poa pratensis* L.). *Agron.* 65:994–995.
8. Berndt, W.L., J.M. Vargas, Jr., A.R. Detweiler, P.E. Rieke, and B. Branham 1987. Black layer formation in highly maintained soils. *Golf Course Management*. June, p. 106.
9. Berndt, W.L., and J.M. Vargas, Jr. 1992. Elemental sulfur lowers redox potential and produces hydrogen sulfide in putting green sand. *Hort. Sci.* 27: 1188–1190.
10. Bird, G.W., and J.M. Vargas, Jr. 1967. Diagnosis and control of nematode problems on turfgrass. *Mich. State. Univ. Ext. Bul.* E-940, 4 pp.
11. Bloom, J.R., and H.B. Couch. 1960. Influence of environment on disease of turfgrasses. 1. Effect of nutrition, pH, and soil moisture on *Rhizoctonia* brown patch. *Phytopathology* 50:532–535.
12. Britton, M.P. 1957. Bluegrass rust studies. *Proc. 1957 Midwest Reg. Turf Conf.*, pp. 78–79.
13. Britton, M.P. 1969. Turfgrass diseases. In *Turfgrass Sci.,* ASA Monogr. Ser., No. 14, pp. 288–335.
14. Brown, G.E., H. Cole, Jr., and R.R. Nelson. 1972. Pathogenicity of *Curvularia* spp. to turfgrass. *Plant Dis. Reptr.* 56(1):59–63.
15. Burpee, L.L., L.M. Kaye, L.G. Goulty, and M.B. Lawton. 1987. Sup-

pression of gray snow mold on creeping bentgrass by an isolate of *Typhula phacorrhiza*. *Plant Dis.* 71: 97–100.

16. Chastagner, G.A. 1984. Necrotic ring spot: Research on a new disease of bluegrass turf and its control in the Pacific Northwest. pp. 94–95 in *Proc. Northwest Turf Conf. 38th*.
17. Chastagner, G.A. 1982. Occurrence of iprodione-tolerant *Fusarium nivale* under field conditions. *Plant Dis.* 66:112–114.
18. Christians, N. 1988. The use of sulfur in turfgrass management. *Landscape Management*. March, pp. 84–88.
19. Crayhay, J.N., P.H. Dernoeden, and N.R. O'Neill. 1988. Growth and pathogenicity of *Leptosphaeria korrae* in bermudagrass. *Plant Dis.* 72:945–949.
20. Christie, J.R. 1959. Plant nematodes—Their biotics and control. Univ. Fla. Agric. Exp. Sta., Gainesville, p. 256.
21. Clarke, B.B., and A.B. Gould. 1993. Turfgrass patch diseases caused by ectotrophic root-infecting fungi. APS Press. Minneapolis, MN, p. 170.
22. Colbaugh, P.F., and R.M. Endo. 1974. Drought stress: An important factor stimulating the development of *Helminthosporium sativum* on Kentucky bluegrass. *Proc. Second Int. Turfgrass Res. Conf.*, pp. 328–334.
23. Cole, H., B. Taylor, and J. Duich. 1968. Evidence of differing tolerance to fungicides among isolates of *Sclerotinia homoeocarpa*. *Phytopathology* 56:683–686.
24. Couch, H.B., and E.R. Bedford. 1966. Fusarium blight of turfgrasses. *Phytopathology* 56:781–786.
25. Couch, H.B., and J.R. Bloom. 1960. Influence of environment on diseases of turfgrasses. II. Effect of nutrition, pH, and soil moisture on *Sclerotinia* dollar spot. *Phytopathology* 50:761–763.
26. Coursen, B.W., and W.R. Jenkins. 1958. Host-parasite relationships of the pin nematode, *Paratylenchus projectus,* on tobacco and on fescue (Abstr.). *Phytopathology* 48:460.
27. Coursen, B.W., R.W. Rhode, and W.R. Jenkins. 1958. Additions to the host lists of the nematodes, *Paratylenchus projectus* and *Trichodorus christiei*. *Plant Dis. Reptr.* 42:456–460.
28. Courtney, W. D., and H. B. Howell. 1952. Investigations on bentgrass nematode *Anguina agrostis* (Steinbuch 1799) Filipjev 1936. *Plant Dis. Reptr.* 36:75–83.
29. Danneberger, T.K., and J.M. Vargas, Jr. 1982. Systemic activity of iprodione in *Poa annua* and post infection activity for *Dreschlera sorokiniana* leaf spot management. *Plant Dis.* 66:914–915.
30. Danneberger, T.K., J.M. Vargas, Jr, and A.L. Jones. 1984. A model for weather-based forecasting of anthracnose on annual bluegrass. *Phytopathology* 74:448–451.
31. Danneberger, T.K., J.M. Vargas, Jr., P.E. Rieke, and J.R. Street. 1985.

Anthracnose development on annual bluegrass in response to nitrogen carriers and fungicide application. *Agron. J.* 75:35–38.

32. Dernoeden, P.H. 1987. Management of take-all patch of creeping bentgrass with nitrogen, sulfur, and phenyl mercury acetate. *Plant Dis.* 71:226–229.
33. Dernoeden, P.H., and N. Jackson. 1980. Infection and mycelial colonization of gramineous host by *Sclerophthora macrospora. Phytopathology* 70: 1009–1013.
34. Dernoeden, P.H., J.N. Crayhay, and D.B. Davis .1991. Spring dead spot and bermudagrass quality as influenced by nitrogen source and potassium. *Crop. Sci.* 31:1674–1680.
35. Detweiler, A.R., J. M. Vargas, Jr., and T.K. Danneberger. 1983. Resistance of *Sclerotinia homoeocarpa* to iprodione and benomyl. *Plant Dis.* 67: 627–630.
36. Detweiler, A.R., J.M. Vargas, Jr., and W.L. Brendt. 1989. Resistance of *Colletotrichum graminicola* to benomyl. *Proceedings Sixth International Turfgrass Conference*. 6:359–362.
37. Davidse, L. C. 1973. Antimitotic activity of methyl benzimidazole-2-yl carbamate (MBC) in *Aspergillus nidulans. Pestic. Biochem. Physiol.* 3:317–325.
38. Davis, D.B., and P.H. Dernoeden. 1991. Summer patch and Kentucky bluegrass quality as influenced by cultural practices. *Agron. J.* 83:670–677.
39. DiEdwardo, A.A. 1963. Pathogenicity and host-parasite relationships of nematodes on turf in Florida. *Fla. Agric. Exp. Sta. Ann. Rep.*, p. 109.
40. DiEdwardo, A. A., and V.G. Perry. 1964. *Heterodera leuceilyma* N.SP. *(Nemata: Heteroderidae)* a severe pathogen of St. Augustinegrass in Florida. *Fla. Agric. Exp. Sta. Bull.*, No. 687, pp. 1–35.
41. Distribution of plant-parasitic nematodes in the South. 1960. Report of the USDA Regional Project S-19. *Southern Cooperative Ser. Bull.*, No. 74.
42. Duell, R.W., and R.M. Schmit. 1974. Grass varieties for roadsides. *Proc. Second Int. Turfgrass Res. Conf.*, pp. 541–550.
43. Elliott, M.L. 1991. Determination of an etiological agent of Bermudagrass decline. *Phytopathology* 81:1380–1384.
44. Elliott, M.L., A.K. Hagan, and J.M. Mullen. 1993. Association of *Gaeumannomyces graminis var. graminis* with a St. Augustinegrass disease. *Plant Dis.* 77:206–209.
45. Ellis, M.B. 1971. *Dematiaceous Hyphomycetes*. Commonwealth Mycological Institute, Kew, Surrey, England, 607 pp.
46. Ellis, M.B. 1976. More *Dematiaceous Hyphomycetes*. Commonwealth Mycological Institute, Kew, Surrey, England, 507 pp.
47. Endo, R.M. 1961. Turfgrass diseases in Southern California. *Plant Dis. Reptr.* 45:869-873; see also *Calif. Turfgrass Cult.* 10(3):22.
48. Endo, R.M. 1963. Influence of temperature on rate of growth of five

fungus pathogens of turfgrass and on rate of disease spread. *Phytopathology* 53:857–861.

49. Endo, R.M., and I. Malca. 1965. Morphological and cytohistological responses of primary roots of bentgrass to *Sclerotinia homoeocarpa* and D-galactose. *Phytopathology* 55:781–789.
50. Endo, R.M., L. Malca, and E.M. Krausman. 1964. Degeneration of the apical meristem and apex of bentgrass roots by a fungal toxin. *Phytopathology* 54:1175–1176.
51. Endo, R.M., H.D. Ohr, and E.M. Krausman. 1985. *Leptosphaeria korrae*, a cause of spring dead spot of bermudagrass in California. *Plant Dis.* 69: 235–237.
52. Feder, W.A., and J. Feldmesser. 1957. Additions to the host list of *Radopholus imilia,* the burrowing nematode. *Plant Dis. Reptr.* 41:33.
53. Filer, T.H., Jr. 1965. Damage to turfgrasses caused by cyanogenic compounds produced by *Marasmius oreades,* a fairy ring fungus. *Plant Dis. Reptr.* 49:571–574.
54. Filer, T.H., Jr. 1966. Effect on grass and cereal seedlings of hydrogen cyanide produced by mycelium and sporophores of *Marasmius oreades. Plant Dis. Reptr.* 50:264–266.
55. Filer, T.H., Jr. 1966. Red thread found on bermudagrass. *Plant Dis. Reptr.* 50:525–526.
56. Freeman, T.E. 1964. Influence of nitrogen on severity of *Pyricularia grisea* infection of St. Augustinegrass. *Phytopathology* 53:187–189.
57. Freeman, T.E. 1965. Rust of *Zoysia* spp. in Florida. *Plant Dis. Reptr.* 49:382.
58. Freeman, T.E. 1969. Diseases of turfgrasses in warm-humid regions. *Proc. First Int. Turfgrass Res. Conf.*, pp. 340–345.
59. Gaskin, T.A. 1966. Evidence for physiologic races of stripe smut *(Ustilago striiformis*) attacking Kentucky bluegrass. *Plant Dis. Reptr.* 40:430–431.
60. Good, J.M., A.E. Steele, and T.J. Ratcliffe. 1959. Occurrence of plant parasitic nematodes in Georgia turf nurseries. *Plant Dis. Reptr.* 43:236–238.
61. Goodman, D.M. and L.L. Burpee. 1991. Biological control of dollar spot of creeping bentgrass. *Phytopath.* 81:1438–1446.
62. Goss, R.L. 1969. Some inter-relationships between nutrition and turfgrass diseases. *Proc. First Int. Turfgrass Res. Conf.*, pp. 351–361.
63. Goss, R.L., and C.J. Gould. 1964. Nitrogen, phosphorus, sulfur and pests. *Proc. Mich. Turfgrass Conf.* 3:75–77.
64. Heald, C.M. 1969. Histopathology of 'Tifdwarf' bermudagrass infected with *Meloidogyne graminis* (Sledge and Golden) Whitehead. 1. *Nematol.* 1(1):9–10.
65. Heald, C.M., and G.W. Burton. 1968. Effect of organic and inorganic nitrogen on nematode populations on turf. *Plant Dis. Reptr.* 52:46–48.

66. Heald, C.M., and V.G. Perry. 1969. Nematodes and other pests. In *Turfgrass Sci.,* ASA Monogr. Ser., No. 140, pp. 358–369.
67. Healy, M.J., M.P. Britton, and J.D. Butler. 1965. Stripe smut damage on 'Pennlu' creeping bentgrass. *Plant Dis. Reptr.* 49:710.
68. Hearn, J.H. 1943. *Rhizoctonia solani* Kuhn and the brown patch disease of grass. *Proc. Texas Acad. Sci.* 26:41–42.
69. Hendrix, F.F., Jr., W.A. Campbell, and J.B. Moncrief. 1970. *Pythium* species associated with golf turfgrasses in the south and southeast. *Plant Dis. Reptr.* 54(5):419–421.
70. Hodges, C.F. 1969. Morphological differences in Kentucky bluegrass infected by stripe smut. *Plant Dis. Reptr.* 53:967–968.
71. Hodges, C.F., and M.P. Britton. 1969. Infection of Merion bluegrass, *Poa pratensis,* by stripe smut, *Ustilago striiformis. Phytopathology* 59:301–304.
72. Hodges, F., and J.P. Madsen. 1978. The competitive and synergistic interactions of *Drechslera sorokiniana* and *Curvularia geniculata* on leaf spot development on *Poa pratensis. Can. 1. Bot.* 56(10):1240–1247.
73. Hodges, C.F., and D.P. Taylor. 1965. Morphology and pathological histology of bentgrass roots infected by a member of the *Meloidogyne incognita* group. *Nematologica* 11:40.
74. Hodges, C.F., and D.P. Taylor. 1966. Host-parasite interactions of a root-knot nematode and creeping bentgrass, *Agrostis palustris. Phytopathology* 56:88–91
75. Howard, F.L. 1953. *Helminthosporium-Curvularia* blights of turf and their cure. *Golf Course Reptr.* 21(2):5–9.
76. Howard, F.L., and M.E. Davies. 1953. *Curvularia* 'fading out' of turfgrasses (Abstr.). *Phytopathology* 43:109.
77. Jackson, N. 1970. Evaluation of some chemicals for control of stripe smut in Kentucky bluegrass turf. *Plant Dis. Reptr.* 54:168–170.
78. Jackson, N., and J.M. Fenstermacher. 1974. Fungicidal control of stripe smut and melting-out with consequent maintenance of sward density in 'Merion' Kentucky bluegrass turf. *Plant Dis. Reptr.* 48(6):573–576.
79. Jackson, N., and P.H. Dernoeden. 1980. *Sclerophthora macrospora:* The incitant of yellow tuft of turfgrasses. *Plant Diseases* 64:915-916.
80. Joyner, B.C., R.E. Partyka, and P.O. Larsen. 1977. *Rhizoctonia* brown patch of Kentucky bluegrass. *Plant Dis. Reptr.* 61:749–752.
81. Kackley, K.E., A.P. Grybauskas, P.H. Dernoeden, and R.L. Hill. 1990. Role of drought stress in the development of summer patch in field inoculated Kentucky bluegrass. *Phytopathology* 80:655–658.
82. Kackley, K.E., A.P. Grybauskas, R.L. Hill, and P.H. Dernoeden. 1990. Influence of temperature-soil water status interactions on the development of summer patch in *Poa* spp. *Phytopathology* 80:650–655.
83. Kelsheimer, E.G., and A.J. Overman. 1953. Notes on some ectoparasitic nematodes found attacking lawns in the Tampa Bay area. *Proc. Fla. State Hort.* 66:301–303.

84. Kozelnicky, G.M., and W.N. Garrett. 1966. The occurrence of zoysia rust in Georgia. *Plant Dis. Reptr.* 50:839.
85. Kreitlow, K.W., F.V. Juska, and R.T. Haard. 1965. A rust on *Zoysia japonica* new to North America. *Plant Dis. Reptr.* 49:185–186.
86. Lacy, M.L., and J.M. Vargas, Jr. 1977. Resistance of Fungi to Fungicides, pp. 239-256. In H. vonR.Wegler (ed.), *Chemie de Pflanzenschutz- und Schadlings bekiimpfungsmittel,* Vol. 4. SpringerVerlag, Berlin.
87. Landschoot, P.J., and N. Jackson. 1989. *Magnaporthe poae* sp. nov., a hyphopodiate fungus with a *Phialophora* anamorph from grass roots in the United States. *Mycol. Res.* 93:59–62.
88. Landschoot, P.J., and B.F. Hoyland. 1992. Gray leaf spot of perennial ryegrass turf in Pennsylvania. *Plant Dis.* 1280–1282.
89. Larsen, P.O. 1979. Leaf blight and crown rot of Toronto creeping bentgrass. *Greens Sect. Rec.* 17:5–7.
90. Laughlin, C.W., and J.M. Vargas, Jr. 1970. Bentgrass nematode control. *Fungicide and Nematicide Tests* 26:152.
91. Laughlin, C.W., and J.M. Vargas, Jr. 1972. Influence of benomyl on the control of *Tylenchorhynchus dubius* with selected nonfumigant nematicides. *Plant Dis. Reptr.* 56:546–548.
92. Laughlin, C.W., and J.M. Vargas, Jr. 1972. Pathogenic potential of *Tylenchorhynchus dubius* on selected turfgrasses. *J. Nematol.* 4:277–289.
93. Laughlin, C.W., and J.M. Vargas, Jr. 1972. In *Fungicide-Nematicide Tests* 27:164–165.
94. Lebeau, J.B. 1964. Control of snow mold by regulating winter soil temperature. *Phytopathology* 54:693–696.
95. Lebeau, J.B., and C.E. Logsdon. 1958. Snow mold of forage crops in Alaska and Yukon. *Phytopathology* 48:148–150.
96. Lucas, L.T. 1976. Chemical control of spring dead spot. *Proc. Am. Phytopathol. Soc.* 3:339.
97. McCarty, L.B., and L.T. Lucas. 1989. *Gaeumannomyces graminis* associated with spring dead spot of bermudagrass in southeastern United States. *Plant Dis.* 73:933–937.
98. McCoy, N.L., R.W. Toler, and J. Amador. 1969. St. Augustine Decline (SAD)—A virus disease of St. Augustinegrass. *Plant Dis. Reptr.* 53:955–958.
99. Madison, J.H. 1971. *Practical Turfgrass Management.* Van Nostrand-Reinhold, New York, p. 466.
100. Mai, W.F., H.W. Crittenden, and W.R. Jenkins. 1960. Distribution of stylet-bearing nematodes in the northeastern United States. Report of the Technical Committee of the NE Regional Project, NE-34. *N.J. Agric. Exp. Sta. Bull.,* p. 795.
101. Martin, S.B. 1984. Characterization and pathogenicity of *Rhizoctonia* species and binucleate *Rhizoctonia*-like fungi from turfgrasses in North Carolina. *Phytopathology* 74:170–175.

102. Meyer, W.A., and A.J. Turgeon. 1975. Control of red leaf spot on 'Toronto' creeping bentgrass. *Plant Dis. Reptr.* 59(8):642–645.
103. Mirza, K., and V.G. Perry. 1967. Histopathology of St. Augustinegrass roots infected by *Hypsoperine graminis* Sledge and Golden. *Nematologica* 13:146–147.
104. Monteith, J., Jr., and A.S. Dahl. 1932. Turf diseases and their control. *USGA Green Sect. Bull.* 12(4):87–187.
105. Mueller, W.C., N. Jackson, and J.M. Fenstermacher. 1974. Occurrence of *Sclerophthora macrospora* in turfgrass affected with yellow tuft. *Plant Dis. Reptr.* 58:848–850.
106. Nelson, E.B. and C.M. Craft. 1991. Introduction and establishment of strains of *Enterobacter cloacae* in golf course turf for biological control of dollar spot. *Plant Dis.* 75: 510–514.
107. Nelson, E.B., and C.M. Craft. 1992. Suppression of dollar spot on creeping bentgrass and annual bluegrass turf with compost-amended topdressings. *Plant Dis.* 76: 954–958.
108. Nutter, F.W., H. Cole, Jr., and R.D. Schein. 1983. Disease forecasting system for warm weather Pythium blight of turfgrass. *Plant Dis.* 67:1126–1128.
109. Parris, G.K. 1957. Screening Mississippi soils for plant parasitic nematodes. *Plant Dis. Reptr.* 41:705.
110. Perry, V.G. 1953. The awl nematode, *Dolichodorus heterocephalus,* a devastating plant parasite. *Proc. Helminthol. Soc. Wash.* 20(1):21–27.
111. Perry, V.G. 1958. A disease of Kentucky bluegrass incited by certain spiral nematodes. *Phytopathology* 48:397.
112. Perry, V.G., H.M. Darling, and G. Thorne. 1959. Anatomy, taxonomy, and control of certain spiral nematodes attacking bluegrass in Wisconsin. *Univ. Wis. Agric. Exp. Sta. Res. Bull.,* No. 207.
113. Perry, V.G., and K.M. Maur. 1968. The pseudo-knot nematode of turf grasses. *Florida Turf* 1(1):3–8.
114. Powell, W.M. 1964. The occurrence of *Tylenchorhynchus maximus* in Georgia. *Plant Dis. Reptr.* 48:70–71.
115. Remsberg, R.E. 1940. The snow molds of grains and grasses caused by *Typhula itoana* and *Typhula idahoensis. Phytopathology* 30:178–180.
116. Rhoades, H.L. 1962. Effects of sting and stubby-root nematodes on St. Augustinegrass. *Plant Dis. Reptr.* 46:424–427.
117. Rhoades, H.L. 1962. Effects of sting and stubby-root nematodes on St. Augustinegrass. *Proc. Univ. Fla. Turf-Grass Manage. Conf.* 10:28–29.
118. Rhoades, H.L. 1965. Parasitism and pathogenicity of *Trichodorus proximus* to St. Augustinegrass. *Plant Dis. Reptr.* 49:259–262.
119. Riffle, J.W. 1964. Root-knot nematode on African Bermuda grass in New Mexico. *Plant Dis. Reptr.* 48:964–965.
120. Riggs, R.D., J.L. Dale, and M.L. Hamblen. 1962. Reaction on bermudagrass varieties and lines to root-knot nematodes. *Phytopathology* 52:587–588.

121. Rivers, J.E., and A.A. DiEdwardo. 1963. The effect of sting (*Belonolaimus longicaudatus*) and lance (*Hoplolaimus coronatus*) nematodes on bermudagrass. *Proc. Univ. Fla. Turf-Grass Manage. Conf.* 11:150–151.
122. Roberts, D.L., J.M. Vargas, Jr., R. Detweiler, K.K. Baker, and G.R. Hooper. 1981. Association of a bacterium with a disease of Toronto creeping bentgrass. *Plant Dis.* 65: 1114–1116.
123. Roberts, D.L., J.M. Vargas, Jr., R. Detweiler, and K.K. Baker. 1982. System suppression with oxytetracycline of a Toronto creeping bentgrass disease of presumed etiology. *Plant Dis.* 66: 804–806.
124. Rowell, J.B. 1951. Observation on the pathogenicity of *Rhizoctonia solani* on bentgrass. *Plant Dis. Reptr.* 35:240–242.
125. Sanders, P.L. 1978. Uptake, translocation, and efficacy of triadimefon in control of turfgrass pathogens. *Phytopathology* 68:1482–1487.
126. Sanders, P.L., L.L. Burpee, H. Cole, Jr., and J.M. Duich. 1978. Control of fungal pathogens of turfgrass with the experimental iprodione fungicide RP 26019. *Plant Dis. Reptr.* 62:549–553.
127. Sanders P.L., 1984. Failure of metalaxyl to control Pythium blight of turfgrass in Pennsylvania. *Plant Dis.* 68:776–777.
128. Sanders, P.L., W.J. Houser, P.J. Parish, and H. Cole. 1985. Reduced rate fungicide mixtures to delay fungicide resistance and to control selected turfgrass diseases. *Plant Dis.* 69:939–943.
129. Shearman, R.C., and P.E. Rieke. 1972. How to interpert soil test results for turfgrasses. *42nd Ann. Mich. Turfgrass Conf. Proc.* 1:93–96.
130. Shurtleff, M.C., T.W. Fermanian, and R. Randell. 1987. *Controlling Turfgrass Pests*. Prentice-Hall, Inc. Englewood, N. J.
131. Smiley, R.W., P.H. Dernoeden, and B.B. Clark. 1992. *Compendium of Turfgrass Diseases.* APS Press. Minneapolis, MN (2nd ed.), p. 98
132. Smiley, R.W., and M. Craven-Fowler. 1984. *Leptosphaeria korrae* and *Phialophora graminicola* associated with Fusarium blight syndrome of *Poa pratensis* in New York. *Plant Dis.* 68:440–442.
133. Smith, J.D. 1954. A disease of *Poa annua*. *J. Sports Turf Res. Inst.* 8(29):344–353.
134. Smith, J.D. 1955. Fungi and turf disease. Dollar spot disease. *Sports Turf Res. Inst.* 9:34–59.
135. Smith, J.D. 1959. The effect of lime application on the occurrence of *Fusarium* patch disease on forced *Poa annua* turf. *J. Sports Turf. Res. Inst.* 9(34):467–470.
136. Smith, J.D. 1974. Snow molds of turfgrasses in Saskatchewan. *Proc. Second Int. Turfgrass Res.* Conf., pp. 313–324.
137. Smith, J.D., N. Jackson, and A.R. Woolhouse. 1989. *Fungal Diseases of Amenity Turf Grasses*. E. & F.N. Spoon, London.
138. Smithson, H., L.B. Loring, and H.J. Jensen. 1963. The "grass-root-gall nematode" found on annual bluegrass roots in Oregon. *Plant Dis. Reptr.* 47:440–441.

139. Stienstra, W.C. 1978. Snow molds on Minnesota golf greens. *Third Int. Turfgrass Res. Conf.* 3:271–274.
140. Sturgeon, R.V., and K.E. Jackson. 1978. Role of soil fungicides and nematicides in maintaining bermudagrass *(Cynodon dactylon)* and creeping bentgrass (*Agrostis palustris*) turf. *Proc. Third Int. Turfgrass Res. Conf.* 3:293–300.
141. Tarjan, A.C., and S.W. Hart. 1955. Occurrence of yellow tuft of bentgrass in Rhode Island. *Plant Dis. Reptr.* 39: 185.
142. Tisserat, N.A., J.C. Pair, and A. Nus. 1989. *Ophiosphaerella herpotricha*, a cause of spring dead spot of bermudagrass in Kansas. *Plant Dis.* 73:933–937.
143. Toler, R.W., and W.J. Walia. 1976. *Turf Diseases TAES Prog. Rep. RP-3375C,* 30 pp.
144. Troll, J., and R.A. Rohde. 1965. Pathogenicity of the nematodes *Pratylenchus penetrans* and *Tylenchorhynchus claytoni* on turfgrasses (Abstr.). *Phytopathology* 56:1285.
145. Troll, J., and A.C. Taoan. 1954. Widespread occurrence of root parasitic nematodes in golf course greens in Rhode Island. *Plant Dis. Reptr.* 38:342–344.
146. Turgeon, A.J. 1985. *Turfgrass Management*. Reston Publishing Company, Inc. Reston, VA. 2nd Ed.
147. Turgeon, A.J., and J.M. Vargas, Jr. 1980. An approach to turfgrass cultivar evaluation. *Proc. Third Int. Turfgrass Conf.* 3:19–30.
148. Vaartnous, H., and C.R. Elliott. 1969. Snowmolds on lawns and lawngrasses in northwest Canada. *Plant Dis. Reptr.* 53:891–894.
149. Van der Plank, J.E. 1968. *Plant Diseases: Epidemics and Control*. Academic Press, New York.
150. Vargas, J.M., Jr. 1973. A benzimidazole resistant strain of *Erysiphe graminis, Phytopath.* 63:1366–1368.
151. Vargas, J.M., Jr. 1976. Disease poses threat to annual bluegrass. *Golf Superintendent* 44:42–45.
152. Vargas, J.M., Jr., D. Roberts, T.K. Danneberger, M. Otto, and R. Detweiler 1989. Biological management of turfgrass pests and the use of prediction models for more accurate pesticide applications. in Leslie, A.R. and R.L. Metcalf (Ed.) *Integrated Pest Management for Turfgrasses and Ornamentals*. 1989-625-030. U.S. Environmental Agency, Washington D.C., pp. 121–141
153. Vargas, J.M., Jr., J.B. Beard, and K.T. Payne. 1972. The comparative incidence of Typhula blight and Fusarium patch on 56 Kentucky bluegrass cultivars. *Plant Dis. Reptr.* 56:32–34.
154. Vargas, J.M., Jr., and R. Detweiler. 1976. Anthracnose. *46th Ann. Mich. Turfgrass Conf. Proc.* 5:27–28.
155. Vargas, J.M., Jr., R. Detweiler, and J. Hyde. 1977. Anthracnose fertility-fungicide interactions. *Mich. Turfgrass Conf. Proc.* 6:3–12.
156. Vargas, J.M., Jr., K.T. Payne, A.J. Turgeon, and R. Detweiler. 1980.

Turfgrass disease resistance-selection, development and use. *Proc. Turfgrass Disease Symp.*, Columbus, Ohio, June 1979.

157. Vargas, J.M., Jr., and A.J. Turgeon. 1975. Translocation of C-14 labelled chloroneb in three turfgrass species. *Can. J. Plant Sci.* 55:85–88.
158. Vargas, J.M., Jr., and A.J. Turgeon. 1978. The principles of blending Kentucky bluegrass cultivars for disease resistance. *Proc. Third Int. Turfgrass Res. Conf.* 3:45–54.
159. Vargas, J.M., Jr., H.C. Young, and E.E. Saari. 1967. The effect of light and temperature on urediospore germination infection and disease development of *Puccinia cynodontis,* and isolation of pathogenic races. *Phytopathology* 57:405–409.
160. Vargas, J.M., Jr., R. Golembiewski, and A.R. Detweiler. 1992. Dollar spot resistance to the DMI fungicides. *Golf Course Management*. March, pp. 56–60.
161. Vargas, J.M., Jr., and A.R. Detweiler. 1985. Anthracnose on *Poa annua*: the pathogenicity of *Colletotrichum graminicola*. *Proceedings Fifth International Turfgrass Conference* 5:637–640.
162. Walter, J., and A.M. Smith. 1972. *Leptosphaeria narmari* and *L. korrae* spp. nov., two long-spored pathogens of grasses in Australia. *Trans. Br. Mycol. Soc.* 58:459–466.
163. Wadsworth, D.F., and H.C. Young, Jr. 1960. Spring dead spot of bermudagrass. *Plant Dis. Reptr.* 44:516–518.
164. Weiling, J.L., S.G. Jensen, and R.I. Hamilton. 1957. *Helminthosporium sativum,* a destructive pathogen of bluegrass. *Phytopathology* 47:744–746.
165. Wilcoxen, S.N. 1976. The complex nature of spring dead spot. *Golf Superintendent* 44:36–38.
166. Wilkinson, J.F., and D.T. Duff. 1972. Rooting of *Poa annua* L., Poa pratensis L. and *Agrostis palustris* Huds. at three soil bulk densities. *Agron. J.* 64:66–68.

Glossary

Acervulus (pl. acervuli)	Erumpent, saucer-shaped, cushionlike fruiting body of a fungus bearing conidiophores, conidia, and sometimes setae.
Acid soil	A soil with a pH value of less than 7.0.
Active ingredient (AI)	The amount of actual pesticide in a formulation.
Aerobic	Relating to the growth of organisms in the presence of molecular oxygen.
Alkaline soil	A soil with a pH value greater than 7.0.
Alternate host	One of two different types of plants on which some pathogens must develop to complete their life cycle.
Anaerobic	Relating to biological processes that occur in the absence of molecular oxygen.
Annual	A plant that completes its life cycle from seed to seed in 1 year.
Antagonism	The phenomenon that results in a depression of activity of an organism when two or more occur in close proximity.
Anthracnose	A leaf disease caused by fungi that produce spores in an acervulus.
Antibiotic	A substance produced by one microorganism that inhibits or kills other microorganisms.
Bactericide	A chemical that kills bacteria; usually an antibiotic.
Bacterium (pl. bacteria)	A microscopic, one-celled, nongreen organism that reproduces by dividing in half. Some bacteria attack plants, causing crown gall, wilts, cankers, rots, leaf spots, blights, and other plant diseases.
Basidiomycetes	A group of fungi that produce sexual spores called basidiospores.

Basidiospore	A sexual fungal spore produced by basidiomycetes.
Basidium	A fungal structure on which basidiospores are produced.
Biennial	A plant that completes its life cycle in 2 years. The first year it produces leaves and stores food; the second year it produces fruits and seeds.
Biocide	A chemical that is toxic to many organisms, including both plants and animals.
Biological control	Control of pests by means of living organisms such as predators, parasites, and disease-producing organisms.
Biological management	Techniques for managing pests by use of predators, parasites, and disease-producing organisms.
Blade	The flattened portion of a grass leaf located above the sheath; the portion of the leaf that grows away from the stem.
Blend	A combination of two or more cultivars of a single turfgrass species.
Carbohydrate	Compound of carbon, hydrogen, and oxygen, as in sugar, starch, and cellulose.
Causal agent	The organism or abiotic factor that produces a given disease.
Chlamydospore	Thick-walled asexual spore formed via modification of a hyphal or conidial cell.
Chloroplast	A structure in plant cells that contains chlorophyll and is the site of photosynthesis.
Chlorosis	Yellowing of normally green plant tissue resulting from lack of chlorophyll. Chlorosis is a common symptom of disease.
Compost	A mixture of organic residues and soil that has been piled, moistened, and allowed to decompose.
Conidiophore	A specialized fungal hypha on which asexual spores (conidia) develop.
Conidium	An asexual fungal spore formed on a conidiophore.
Contact pesticide	A pesticide that kills on contact.
Cool-season turfgrass	Turfgrass species adapted to favorable or

optimal growth during cool portions (60 to 70°F [15.5°–21.1°C]) of the growing season.

Core aerification A method of turf cultivation in which soil cores are removed by hollow tines or spoons to control soil compaction and to aid in the penetration and distribution of pesticides and water.

Cultivar An assemblage of cultivated plants distinguished by any characters (morphological, physiological, cytological, etc.) which, when reproduced sexually or asexually, retain their distinguishing features; a cultivated variety.

Cultivation Applied to turf, cultivation refers to working of the soil and/or thatch without destruction of the turf; e.g., coring, slicing, spiking, or other means.

Cuticle A thin waxy layer on the outer wall of epidermal cells.

Cyst In nematodes, a structure on root that may contain eggs.

Cytoplasm Living substances in a cell except for the nucleus.

Damping-off Decay of seeds in the soil or young seedlings before or after emergence. Most evident in young seedlings that suddenly wilt, topple over, and die from rot at the stem base.

Dermal Pertaining to the skin.

Dermal toxicity Ability of a compound, when absorbed through the skin of animals, to produce symptoms of poisoning.

Desiccation Drying out.

Dethatching Removing excessive thatch from a turfgrass area by hand raking or with various types of machinery equipped with vertical knives or tines.

Disease Abnormal development of a living organism that interferes with its normal structure and reduces its economic or aesthetic value.

Disease cycle Stages involved in the development of a disease.

Disease tolerance In plants, the ability of a host to survive and produce satisfactory yields at a level of infec-

	tion that causes economic loss to other cultivars or types of the same host plant.
Disease triangle	A concept describing the simultaneous occurrence of a virulent pathogen, a susceptible host, and a favorable environment such that a disease may develop.
Dissemination	In relation to plant diseases, spread of inoculum.
DMI	Demethylation inhibiting fungicides. A group of fungicides that prevent demethylation processes from occurring as fungi try to synthesize ergosterol from squaline for cell membrane formation. (Fenarimol, propiconazole, triadimefon.)
Dormant	Not actively growing (as seeds or other plant organs) due to internal causes.
Drought	An extended period of dry weather that usually causes stress to turf.
Ectoparasite	A parasite that feeds on the outside of a host (nematodes).
Egg	In nematodes, the first stage of the life cycle.
Endodermis	A layer of cells that surrounds the vascular tissues of plant roots.
Endoparasite	A parasite that enters and feeds inside a host (nematode).
Endophyte	A fungus living within another plant, which protects the plant from foliar insects.
Enzyme	A protein that catalyzes a specific biochemical reaction.
Epidemic	A rapidly developing and widespread occurrence of a disease.
Epidermis	The outer layer of cells on all plant parts.
Exudates	Liquid discharge from plant tissue.
Facultative parasite	An organism that spends most of its life as a saprophyte but may become parasitic when environmental conditions are favorable.
Facultative saprophyte	An organism that spends most of its life as a parasite, but may live as a saprophyte for short periods of time.
Field capacity	The amount of moisture remaining in the soil after gravitational moisture has drained.

Flagellum	A whiplike structure on bacterial cells or zoospores that functions in movement.
Flowable	A finely ground, wettable powder formulation that is sold as a thick suspension in a liquid.
Formulation	The form in which a pesticide is offered for sale to the user (as emulsifiable concentrate, wettable powder, granule, dust, oil solution, etc.). Includes both the active and inert ingredients.
Fruiting body	A fungal structure in which spores are produced.
Fumigant	A pesticide that, as a gas, kills destructive microorganisms, animals, and plants.
Fumigation	The application of a fumigant to an area.
Fungicide	A pesticide used to prevent, destroy, repel, or mitigate fungal infections.
Fungistat	A pesticide that inhibits the growth of fungi but does not kill them.
Fungus	An organism composed of hyphae and lacking chlorophyll which is saprophytic and/or parasitic.
Gall	An overgrowth produced on a plant.
Gene	The hereditary unit; a segment of DNA coding for a specific protein.
Genus (pl. **genera**)	A more or less closely related and definable group of plants or animals comprising one or more species.
Germ tube	The growth of hyphae from a germinating fungus spore.
Germination	The beginning of growth of a spore, or seed sprouting.
Granules	A formulation in which the pesticide is attached to particles of an inert carrier such as clay or ground corncobs.
Haustorium (pl. **haustoria**)	A specialized hypha of a fungus that penetrates a host plant, makes intimate contact with the host cell's protoplast, and absorbs food. Powdery mildews, downy mildews, and rusts produce haustoria.
Herbicide	A chemical used for killing weeds; a type of pesticide.

Host	Any plant or animal that is invaded by a parasite and from which the parasite obtains its food.
Hyaline	Transparent; without color.
Hydathodes	Structures in which fluids move from the interior to the surface of the leaf.
Hypersensitivity	A response by plants to invasion by certain pathogens which includes rapid dying of local tissue to prevent further advances of the pathogen.
Hypha (pl. hyphae)	Single filament that constitutes the body (mycelium) of a fungus. Hyphae may be divided into cells by cross walls.
Immune	Cannot be infected by a pathogen.
In vitro	Outside the living body (e.g., a host) and in an artificial environment; in culture.
In vivo	In the living body of a plant or animal.
Incubation period	The time between penetration of a host by a pathogen and the appearance of symptoms.
Inert ingredients	Ingredients of a pesticide formulation that are not active, such as water, sugar, dust, wetting and spreading agents, emulsifiers, propellants, etc.
Infection	The early stage of disease caused by a pathogen.
Infectious disease	A disease caused by a pathogen that multiplies and can be transmitted from plant to plant.
Infested	Containing or covered by great numbers of pathogens (insects, nematodes, bacteria, or fungi). Refers to the presence of pathogens in the environment, not to their infection of hosts.
Inoculum	Pathogen or its parts that infect (e.g., fungus spores, mycelium, nematodes, virus particles).
Inoculum	Part of a pathogen that comes in contact with a host and can cause disease.
Integrated pest management (IPM)	Controlling pest populations through chemical, cultural, biological, and genetic methods.
Intercellular	Between the cells.

Internode	Portion of a stem between the nodes or growing points.
Interseeding	Seeding into an established turf.
Intracellular	Within the cell.
IPM	See Integrated pest management.
Lateral shoot	A shoot originating from a vegetative bud in the axil of a leaf or from the node of a stem, rhizome, or stolon.
Layering, soil	Undesirable stratification within the A horizon of a soil profile; can be due to construction design, topdressing with different textured materials, inadequate on-site mixing of soil constituents, or blowing and washing of sand on other soil constituents.
LD_{50}	A measure of toxicity; the dose of a compound that causes death in 50% of the test animals so treated. If a compound has an LD_{50} of 50 mg/kg, it is more toxic than one having an LD_{50} of 500 mg/kg.
Leaching	Movement of a substance downward or out of the soil as the result of water movement.
Leaf spot	A lesion on a leaf.
Lesion	A localized area of diseased tissue.
Localized dry spot	A dry spot of sod amid normal, moist turf that resists rewetting by normal irrigation, associated with a number of factors including thatch and soil fungi. They are most commonly found in high sand content soils.
Meristem (adj. meristematic)	Plant tissue that functions principally in cell division and differentiation.
Microorganism	Minute living organism such as a bacterium, some fungus, some nematode, or protozoan which cannot be seen without the use of a microscope.
Mixture	A combination of two or more species.
Mowing frequency	The number of times a turfgrass community is mowed per week, month, or growing season. (The reciprocal of mowing frequency is mowing interval—the number of days, weeks, etc. between successive mowings.)
Mowing height	The distance above the ground surface at which the turfgrass is cut during mowing.

Mycelium (pl. **mycelia**)	The mass of interwoven filaments (hyphae) that makes up the vegetative body of a fungus.
Mycoplasma	A bacteria-like organism that lacks a rigid cell wall and is variable in shape. Mycoplasmas cause certain plant diseases of the "yellow" type.
Mycorrhiza	A mutualistic symbiotic relationship between a fungus and the roots of a plant.
Necrosis (adj. **necrotic**)	Death, usually accompanied by darkening or discoloration.
Necrotic	Exhibiting varying degrees of dead areas or spots.
Nematicide	A chemical compound that kills nematodes.
Nematodes	Generally microscopic, unsegmented roundworms that usually live free in moist soil, water, or decaying matter, or as parasites of plants and animals.
Node	The joint of a stem; the region of attachment of leaves to a stem.
Obligate Parasite	An organism that can live only on or in living tissue.
Organic matter	Plant or animal material capable of undergoing breakdown and resynthesis (in the soil).
Overseed	To seed onto an existing turf, usually with temporary turfgrass, to provide green, active grass growth during dormancy of the original turf, usually a warm-season turfgrass.
Parasite	An organism (fungus, bacterium, virus, nematode, etc.) that obtains its food from another organism. An obligate parasite is one that can develop only in living tissues (e.g., a virus or plant-parasitic nematode).
Pathogen	An organism or agent capable of causing disease.
Pest	Any insect, mite, nematode, fungus, weed, or other plant or animal that is injurious to the health of plants.
Pesticide	A chemical or mixture of chemicals used to destroy, prevent, mitigate, repel, or attract any animal, plant, or plant disease considered a pest. Examples are insecticides, miticides

	herbicides, fungicides, nematicides, and bactericides.
pH	A measure of acidity and alkalinity within the range of 0 to 14. The midpoint value (7) represents neutrality (neither acid nor alkaline); values below 7 have increasing acidity or hydrogen ion concentrations; and those above 7 have increasing alkalinity or lower hydrogen ion concentrations.
pH, soil	A numerical measure of the acidity or hydrogen ion activity of a soil. A pH of 7 indicates neutrality; above 7 is basic (alkaline), while below 7 is acidic.
Phloem	Food-conducting tissues in plants.
Photosynthesis	Process by which carbohydrates are produced from carbon dioxide, water, and light energy in chlorophyll-containing plants.
Phytotoxic	A pesticide that is injurious or poisonous to plants.
Primary inoculum	Survival structures of a pathogen that cause primary infections.
Protectant	A pesticide used to prevent infection by an organism.
Protoplasm	Living material of a cell.
Pustule	A small blisterlike structure in which some fungi produce spores.
Race (or strain)	A subgroup or biotype within a species of a pathogen that differs in virulence, symptom expression, or host range—limited to certain species, varieties, hybrids, or cultivars—from other races and the rest of the species.
Relative humidity (RH)	The ratio of the quantity of water vapor present in the atmosphere to the quantity that would saturate it at the same temperature.
Residual	The property of a pesticide for persisting after application in amounts sufficient to kill pests for several days to several weeks or even longer.
Residue	The amount of pesticide present following application.
Resistant species	Pests that survive relatively high rates of pesticide application.

Resting spore — A spore, usually thick-walled, that may remain alive in a dormant state for months or years, later germinating and capable of causing infection.

Rhizome — A modified underground rootlike stem that produces roots and leafy shoots.

Saprophyte — An organism that obtains its food from dead organic matter, as opposed to a parasite that feeds on living tissue.

Scalp — To remove an excessive quantity of functioning green leaves at any one mowing; results in a shabby, brown appearance caused by exposing crowns, stolons, dead leaves, and even bare soil.

Sclerotium (pl. sclerotia) — A small, compact, often hard interwoven mass of fungus hyphae, usually more or less spherical or flat, and dark in color.

Septate — In fungi, hyphae with crosswalls.

Seta (pl. setae) — Stiff, hairlike appendage.

Sheath — The tubular, basal portion of the leaf that encloses the stem.

Shoot density — The relative number of shoots per unit area.

Sign — Parts of a pathogen or its products seen on a diseased host plant.

Slicing — A method of turf cultivation in which rotating, flat tines slice intermittently through the turf and the soil.

Slime mold — Primitive organisms whose plasmodium "flows" over low-lying vegetation like an amoeba.

Sod — Plugs, squares, or strips of turfgrass with the adhering soil which is used in vegetative planting.

Sooty mold — A sooty fungal growth on foliage and fruit in the honey dew secreted by insects such as aphids.

sp (pl. spp) — Species; a genus name followed by sp. means that the particular species is undetermined; spp. after a genus name means that several species are being referred to without being named individually.

Species (pl. species) — A group of closely related individuals resem-

	bling one another in certain inherited characteristics.
Spiking	A method of turf cultivation in which solid tines or flat, pointed blades penetrate the turf and soil surface.
Spore	In fungi, the microscopic reproductive unit consisting of one or more cells.
Sporulate	To produce spores.
Spot	Limited, chlorotic or necrotic, circular to oval area on leaves or on other plant parts.
Spot treatment	Application to a restricted or small area.
Sprig	A single turfgrass stem (stolon, rhizome, or tiller), usually with some attached roots, that is used in vegetative propagation.
Sprigging	Vegetative planting by placing stolons, rhizomes, or tillers in furrows or small holes.
Stolon	An elongated horizontal stem (or shoot) that grows above the soil surface and roots at the nodes. It is used to propagate certain grasses.
Stoma (pl. stomata)	A minute opening in the epidermis of a leaf or stem through which gases are exchanged.
Stroma (pl. stromata)	Compact mass of mycelium (with or without host tissue) that supports fruiting bodies.
Stylet	Stiff, slender, hollow feeding organ in plant-parasitic nematodes.
Summer dormancy	The cessation of growth and subsequent death of shoots of perennial plants due to heat and/or moisture stress.
Susceptibility	The inability of a plant to overcome disease.
Susceptible	Lacking the ability to resist the effect of a given pathogen.
Suspension	A mixture in which the pesticide in a solid form (tiny particles) is suspended in water.
Symbiosis	Two or more organisms living together.
Symptom	The visual reactions of an organism to a disease.
Syringe	To spray turf with small amounts of water; usually on a hot, dry, windy day to reduce water loss (transpiration).
Systemic	Pertaining to chemicals or pathogens that spread throughout plants as opposed to remaining localized.

Systemic pesticide	A pesticide that is absorbed by treated plants and translocated (moved) to most tissues.
Teliospore	Thick-walled resting spore of rust and smut fungi that germinates to form a basidium.
Thatch	A tightly intermingled layer of undecomposed or partially decomposed organic residues situated above the soil surface and generally below the green portions of the turf.
Tiller	A lateral stem (or shoot), usually erect, that develops from the central crown.
Toxicity	The capacity of a pesticide to cause harm to a living organism.
Translocation	Movement of foods and wastes within a plant.
Transmission	Spread of pathogens from plant to plant.
Transpiration	The loss of water vapor from plant surfaces.
Urediospore, uredospore, or urediniospore	Binucleate, dikaryotic, asexual rust spore.
Variety	A subdivision of a species where individuals within a species are distinct in form and function from other similar arrays of individuals. (The term cultivar is preferred.)
Vascular	Pertaining to conductive (xylem and phloem) tissue.
Vector	Agent that transmits inoculum.
Vegetative propagation	Asexual propagation using pieces of vegetation, i.e., sprigs or sod pieces.
Virulence	The degree of pathogenicity of a pathogen.
Virulent	Capable of causing a disease.
Virus	A submicroscopic obligate parasite composed of nucleic acid and protein.
Warm-season turfgrass	Turfgrass species adapted to an optimal growth during warm portions (80 to 95°F [26.6° to 32°C] of the growing season.
Water-soaked	Pertaining to plants or lesions that appear wet, dark, and usually sunken and translucent.
Weed	Any plant growing in a place where it is not wanted.
Wet wilt	Wilting of turf in the presence of free soil

	water when evapotranspiration exceeds the ability of roots to take up water.
Wettable powder	A powder formulation containing a wetting agent that causes the powder to form a suspension in water.
Wilt	Loss of rigidity of plant parts due to insufficient water in the plant.
Winter desiccation	The death of leaves or plants by drying during winter dormancy.
Winterkill	Any injury to turfgrass plants that occurs during the winter period.
Witches' broom	Broomlike growth from the dense clustering of branches.
Xylem	Water-conducting tissue in plants.
Yellows	A plant disease characterized by yellowing and stunting of plants.
Zoospore	A swimming spore or swarm spore capable of independent movement; especially in the lower fungi (*Pythium*).

Index

For Product Safety Concerns and Information please contact our
EU representative GPSR@taylorandfrancis.com Taylor & Francis
Verlag GmbH, Kaufingerstraße 24, 80331 München, Germany